go gently

BONNIE WRIGHT

go gently

ACTIONABLE STEPS
TO NURTURE YOURSELF
AND THE PLANET

Illustrations by Michael Haddad

greenfinch

First published in the United States in 2022 by Mariner Books, an Imprint of HarperCollins
First published in Great Britain in 2022 by

Greenfinch
An imprint of Quercus Editions Ltd
Carmelite House
50 Victoria Embankment
London EC4Y 0DZ

An Hachette UK company

Illustrations © Michael Haddad

Recycle symbols pp. 105 © Shutterstock / Askhat Gilyakhov;
recycle symbol p. 104 © Shutterstock / Standard Studio

A CIP catalogue record for this book is available from the British Library

HB ISBN 978-1-52941-741-8
Ebook ISBN 978-1-52941-740-1

10 9 8 7 6 5 4 3 2 1

Printed and bound in Italy by L.E.G.O. SpA

FSC
www.fsc.org
MIX
Paper from
responsible sources
FSC® C023419

Papers used by Greenfinch are from well-managed forests and other responsible sources.

THIS BOOK IS DEDICATED TO
THE FOUR THINGS I WOULD BE
NOTHING WITHOUT:

soil

water

air

light

CONTENTS

INTRODUCTION

My journey into the climate movement has taught me and continues to teach me so much, not only about climate science and nature's systems but also what it means to be human. What responsibility we hold as a keystone species, a species that has a disproportionate effect on other species within the ecosystem. We are coming out of an age where the dominant narrative has been that we are separate from nature and must journey to it, as though nature is something outside of ourselves. Yet nothing and no one exist in isolation. So, we have a responsibility to each other and the future of every species on this incredible rock called Earth. We cannot afford to be complacent.

My fascination with and love for the ocean is what first pulled me into the climate movement. When I saw the ocean, which felt like part of me, polluted with single-use plastics, I was determined to understand why this was happening and what I could do. Through working with Greenpeace, Rainforest Alliance and Waste Watch, the local grassroots group I cofounded, I have had the privilege of meeting people with intersectional approaches and Indigenous knowledge, and those on the front lines of research, all working to tackle the climate crisis. I interviewed some of these people to share their direct experiences in this book. In addition to engaging with these global relationships and communities, I have been implementing changes at home in order to deepen my relationship with the climate movement and to better understand the effects of my actions. It is these small, quiet actions in the intimacy of my home that have allowed me to slow down and listen, to deprogramme social norms of being busy, consuming, productivity, growth, and instead focus on building relationships with the things, people and parts of nature I want to protect, love and seek joy in.

I think a lot of us have preconceived notions about what a climate activist should look or act like – that they should know all the statistics, make grand speeches, commit every moment of their life to the cause. But what the climate really needs is for each of us to show up just as we are. In fact, it was the moments in my journey when I became frustrated and disappointed in myself for not achieving change in this programmed, perfect way that underlined for me the importance of writing this book. There is no perfect way to show up for the planet, each other and yourself. Perfectionism – essentially, ego – distracts us from the urgency with which we need to be addressing the climate crisis. I wrote this book to celebrate imperfect, in-process action.

The more people I have met in the movement, the wider the range of approaches I have seen. People bring their own skill sets, interests and creativity to their application of change. My first approach was a love for the ocean and a fascination with waste and our relationship to consumption. Yet someone else's approach could stem from a love for the potential of urban environments and cooking, which might manifest as being part of a rooftop community garden. The opportunity to take action within the climate crisis is limitless. We don't have to choose one thing, and we don't have to land on it right away. It is a continual inquiry.

I was drawn to the medium of a book as a means to collect and share my thoughts in a tangible way. The physical object itself – a book you can feel in your hands – is a reminder that we can hold and see change. The inspiration behind *go gently* was to show that we can adopt both hard and soft mind-sets, we can *go* forth with action but still foster a loving, *gentle* relationship to the planet, each other and ourselves. I am so excited to be building a community between the pages of this book as we collectively grow through the actions we take.

go learn

A BIG-PICTURE LOOK AT
THE CLIMATE MOVEMENT

Before we dive into at-home practices that support a healthy planet for all creatures, it is essential that we take a look at the systems that govern our planet. This includes those designed by humans and those that pre-exist within Earth's blueprint. We will explore how these systems interact, which of them are broken, which are thriving and which we need to support so that we can mitigate climate breakdown. The ice caps are not *going* to melt – they already are. Sea levels are not *going* to rise – they already are. This is why we need to take a crisis approach to the climate. Fossil fuel companies are not going to put the planet before profit – we as compassionate earthlings need to do that. The time to join this movement and educate ourselves on how to best take action is now.

Many systems built long ago, such as slavery and colonialism, have entrenched effects on people and the planet. However, just as those systems were once imagined by someone, we have the same opportunity to radically imagine new systems or remember ones that have been forgotten, systems that centre people and the planet equally. The climate crisis is a big issue, but we can approach it with both global and local solutions. Lessons about the environment can often be disheartening, yet the more I have directed my gaze to the positive systems that exist on this planet – both natural and human-designed – the more inspired I become by our potential. As climate justice writer Mary Annaïse Heglar so beautifully puts it, 'The thing about climate is that you can be overwhelmed by the complexity of the problem or fall in love with the creativity of solutions'.

My curiosity and inspiration to change my behaviour at home was instigated by a developing knowledge of the climate crisis. Being exposed to these stories and statistics called me to ask, What can I do about it? What part can I play? We are each personally drawn to and affected by elements that we can work to defend on a micro and macro scale. My best advice is, don't choose to focus on one element just because you think you should or because your friend is. Find something that personally speaks to you and sparks your curiosity and joy. We are much more likely to commit to an issue long term if we feel connected to and inspired by it. It might not be the first topic you pick, but that will lead you to the next, until you find the right fit.

CLIMATE CRISIS 101: KEY TOPICS AND TERMINOLOGY

Let's take a deeper dive into the basics of climate change. We can hear lots of technical terms but never really know what they mean in relation to us and our actions. When I was a kid, 'global warming' was the most popular term to describe our changing climate. Warming temperatures are the result of the greenhouse gas effect.

WHAT IS THE GREENHOUSE GAS EFFECT?

This is when greenhouse gases, including carbon dioxide, methane and nitrous oxide, rise up into Earth's atmosphere, where they trap heat, similar to the glass roof of a greenhouse. This trapped heat is unable to escape into space, so it remains in the atmosphere, and the surplus heat disrupts the equilibrium of Earth's environment. We humans have been emitting these gases at an ever-increasing rate by burning fossil fuels, burying and incinerating our rubbish and engaging in degenerative agricultural and industrial systems. The more these gases are released, the more they get trapped in Earth's atmosphere, and the more they heat up both our land and our oceans.

Greenhouse gases rise into the atmosphere and trap heat from the sun. The atmosphere, depicted here as a chain, locks in this heat, making it unable to escape.

The greenhouse gas effect is resulting in:

> Increased temperatures, making it harder to live and work. This will lead to more energy usage to cool our homes.

> Higher water temperatures and polar ice melting, both of which are driving rising sea levels, which will submerge land masses and contaminate our soil and aquifers with salt.

> Bad air quality, which affects human health.

> More frequent extreme weather events.

> A significant decrease in the amount of land habitable by humans. The World Economic Forum predicts that by 2070, this could affect 30 per cent of the population, which would lead to mass migration as people seek asylum in cooler countries.

> A greater need for water to manage the changing temperatures, including drinking water and water for agriculture.

As we know, there is no 'Planet B', so to keep Planet A – Earth – habitable, we need to radically decrease our use of and dependence on fossil fuels, looking instead to renewable energy sources and practices that do not emit greenhouse gases. The good news is, there are so many positive solutions out there that we can get behind. But until humanity collectively puts pressure on world leaders to take action, they will not put the planet before profit.

The old versus the new. The destructive, pollution-creating fossil fuel industry juxtaposed against the clean and thriving renewable energy industry.

FOSSIL FUEL EXTRACTION, AND WHY IT MUST END!

Imagine the world is an overflowing bathtub, the water being fossil fuel products such as plastics, petrol and coal. As we rush to bail out the bathtub with our buckets to manage the mess, we can't forget to also turn the tap off, the tap being the extraction of *more* fossil fuels. It may sound radical, but our work to mitigate the climate crisis will never catch up if we keep extracting more fossil fuels. We must turn the tap off!

So what *are* fossil fuels? Natural gas, coal and oil are made from decomposed plants and animals that have been underground for millions of years. We have been extracting these substances since the 1800s by drilling into our earth to remove liquid or gaseous fossil fuels such as oil and natural gas, and mining to extract solid fossil fuels such as coal. These materials are then burned in fossil fuel power plants, which generate steam to drive turbines, which in turn generate electricity. Gas and oil are used to fuel vehicle engines. Plastics are also made from petrochemicals, which are the chemical products obtained from fossil fuels.

This burning of fossil fuels – as they are processed and used as energy sources – creates an astronomical amount of destructive greenhouse gases, wh. are absorbed by our oceans and atmosphere, warm up our planet and continue to make the climate less hospitable to humans and many other species. As you can gather, it is time to say bye-bye to fossil fuels and welcome our new BFF, renewable energy!

ACTIONABLE STEPS

- ⊘ Call your representatives and advocate for green industries and jobs. Vote and petition against any new fossil fuel extraction development.

- ⊘ Use public transport, cycle, walk, car-pool and minimize driving.

- ⊘ See if your bank invests in fossil fuels. If so, think about switching to a bank or organization that transparently states that they do not invest in fossil fuel companies. I recently did this, and it was an easier process than I thought.

- ⊘ Limit your use of single-use plastic as much as possible.

RENEWABLE ENERGY, TODAY'S AND TOMORROW'S BFF

To save yourself from spiralling into thoughts of doomsday, let's take a moment to appreciate and celebrate renewable energy systems and how they can help save our future on this planet. While humans have had their fair share of bad press, one positive news story is our ingenuity in designing renewable energy systems that harness energy from nature's greatest gifts: wind, sun and water. Fossil fuels come from resources that are extracted, used up and will eventually run out, but energy from the wind, sun and water are considered 'renewable' because the resource is not depleted when it is used. These renewable energy systems sound pretty amazing, but their implementation needs to be carried out in a just, accessible and anticolonial way in order to address the climate crisis for all living creatures. Let's look at the key types of renewable energy:

Solar energy – Solar panels capture radiant energy from sunlight and convert it into heat, electricity or hot water. Over the long term, solar can greatly reduce energy costs for the consumer. Many governments incentivize investment in solar energy by providing rebates and tax credits.

Wind energy – wind turbines, which look like giant fans, can be installed on land or at sea. Wind turns the blades of the turbine around a rotor, which spins a generator and creates electricity. Wind energy is an efficient use of land, has low operating costs and creates jobs.

Wave energy – devices such as buoys and float systems are anchored out at sea. As waves and the movement of water move these devices up and down, the machinery transforms kinetic energy into electricity. The potential of wave energy has not yet been fully explored because it's more complicated than wind energy or solar energy; the machinery is expensive to install and maintain, and there are potential threats to marine life.

Hydropower – I feel it's important to mention hydropower, as it is the most commonly used renewable energy but not the best – it's the BFF we need to unfriend. Hydropower systems – including the 38,000 dams around the world – can require more energy to run than they are able to produce, and some rely on fossil fuels to pump water. They also significantly disrupt water ecosystems and waterways, which negatively affects the animals that live in those waterways.

ACTIONABLE STEPS

- ⊘ Look for and support companies that run on renewable energies.
- ⊘ If you own a business, or work for a business where you can influence the people in charge, think about shifting to renewable energy sources.
- ⊘ Investigate installing solar panels for your home if that is an option for you.
- ⊘ Consider using an energy provider that sources the energy from renewable resources. Many online comparison sites such as Uswitch now allow you to filter by 'green energy suppliers' to make it even easier.

DR DAVID SANTILLO

Senior scientist with the Greenpeace Research Laboratories, based at the University of Exeter in the UK

What is the correlation between human activity and climate change?

Over the last decade, it's become almost impossible for anyone to deny that there is this thing called climate change and that humans play an overwhelming part in it. All of the scientific data out there point to the same facts: that climate change is real, that we can measure the changes physically in the environment, that we can see ice melting, see climates and weather patterns changing, and we know that it is driven mainly by carbon dioxide, and that a lot of the increase in carbon dioxide is coming directly from human activities. I think now anyone who chooses to deny that climate change is an issue, and that it's an issue for us *now*, is really choosing to ignore all of the evidence that's out there. If there's any doubt, then it's a wilful decision not to accept it.

What is one piece of actionable advice you would give?

We've got to get into a position where we're not just thinking, 'I'll do my bit, and I hope everyone else does the same'. Yes, we can, but the people who can make the biggest difference of all are governments and big corporations. We need to also act by taking the arguments and the messages back to them and saying, 'You need to do more; you need to set things up so that we can make the right choices'. Put pressure on your elected officials, put pressure on the retailers that you buy from. People can make themselves heard and make themselves difficult and not feel bad about doing that, because that's the only way, in the end, that change is really driven through at some of these levels.

What instills hope in you?

The fact that we've got all of this science together with all the work that the IPCC [Intergovernmental Panel on Climate Change] and others are doing, the fact that we've got young people who are engaged, we've got schools that are interested in teaching this as part of the curriculum. We've got so many more scientists now coming through university who are working in this field, whether that's ocean science or climate modelling or pollution monitoring. There is just a much, much greater groundswell of understanding of the fact that everything we do has an impact.

WATER, WATER EVERYWHERE, BUT NOT A DROP TO DRINK

I remember reading a headline years ago that was something like 'When Water Will Be More Expensive Than Oil'. The article completely changed my perspective on water. I was aware that our day-to-day lives depended on water, but I underestimated just how much we have been taking from a resource we viewed as limitless and not seeing how our treatment of water would make it a scarce commodity. As I delved deeper, I discovered how inextricably linked water scarcity and the climate crisis are.

Climate-related events can affect water supply as follows:

- ⊙ Natural disasters such as typhoons and tsunamis can destroy water systems and contaminate the water supply.

- ⊙ Rising sea levels can compromise freshwater supplies by making them salty.

- ⊙ Extreme weather events can affect water-cycle patterns and make them less predictable. This limits people's access to water, making many people vulnerable.

- ⊙ Soil erosion and the inability of land to retain water and grow plants (also known as 'desertification') force people to migrate from affected areas. According to the European Commission's World Atlas of Desertification, more than 75 per cent of Earth's land area is already degraded and more than 90 per cent could become degraded by 2050.

- ⊙ Aquifers are threatened. The International Groundwater Resources Assessment Centre explains that the 'increased variability in precipitation and more extreme weather events caused by climate change can lead to longer periods of droughts and floods, which directly affects availability and dependency on groundwater'.

- ⊙ Rising temperatures cause more water to evaporate from the land and oceans, which over time builds a higher level of atmospheric water vapour and leads to more intense rainfall in the future. Some of this rainwater can be polluted with chemicals from agricultural pesticides.

- ⊙ Oil drilling, fracking and fossil fuel pipelines break up wildlife habitats and contaminate water supplies.

Some of the human actions that create these changes include:

- ⊙ Greenhouse gas emissions
- ⊙ Deforestation
- ⊙ Agriculture
- ⊙ Urbanization
- ⊙ Population growth
- ⊙ Pipeline construction

If we continue at the rate we are extracting and burning fossil fuels, engaging in degenerative farming practices and treating our beautiful planet as a limitless resource, we will all become water vulnerable. The only water we will have left will be salinized, contaminated or extremely expensive. This will exacerbate the water apartheid (disadvantaged access to water) that already exists in America and around the globe.

ACTIONABLE STEPS

- ⊘ Respect and conserve water.
- ⊘ Trace the source of your food and try to eliminate links to agriculture that degenerates landscapes and/or uses chemical pesticides.
- ⊘ Protest against pipelines and any new oil-drilling sites.
- ⊘ Support those who have experienced natural disasters or water apartheid through donations or amplification of their communities and the connection to the climate crisis.
- ⊘ If you are able, collect rainwater and use it for watering your plants or for washing things.
- ⊘ Use nontoxic detergents and cleaning products, as the ingredients can make their way into the ocean and pollute the water.
- ⊘ If you need to dispose of a product that is deemed hazardous, such as paint or drain cleaner, check to see if your area has a Household Hazardous Waste special collection day or drop-off.
- ⊘ Research whether your city's drinking water infrastructure is equally available to everyone in the community.

JOANNA SUSTENTO
Climate justice activist

What does 'climate justice' mean to you, and how has your own experience as a survivor of Typhoon Haiyan and losing your family informed the advocacy work you do?

Before Typhoon Haiyan, I didn't really care much about what was happening in the environment. I was aware of the problems, but I didn't have a deeper understanding of how the system is broken. After the disaster, I realized that climate change *is* a human rights issue. I really believe that what our community can put forth in the climate justice conversation is the human face of the climate crisis. When we talk about the climate crisis, it's very focused on climate science. Not a lot of people can relate to that, but when you talk about losing families, losing livelihood, when you talk about years of hard work disappearing in the span of two hours, a lot of people can take that to heart and can really empathize with that.

What is one piece of actionable advice you would give?

It's important that at an early age young people realize what they can contribute to their community. They should always speak up and create that safe and brave space for their fellow youth so that they can be heard. You don't need to be a celebrity. You don't need to be a politician. You don't need to be someone influential or rich to be able to create that impact in your community.

What instills hope in you?

Whenever I'm having a bad day or am grieving from what I've been through, I remind myself that the universe led me here. I keep reminding myself of my 'why', and my why is I just couldn't stop pursuing this advocacy, because I've already experienced the worst. I know how it feels to lose everything, and I don't want other people to go through what I've gone through. I promised myself that at the end of this second life I was given, I would make sure my community wins.

AIR QUALITY: JUST BECAUSE THE SKY IS BLUE DOESN'T MEAN IT'S CLEAN

When I have shared on social media information about policies such as AB-345 in California (which disallows any fracking site from being built or existing within 760m/2,500ft of houses, schools or parks), I have received many responses from people who grew up by such sites and are now suffering from health issues. Often the most polluted areas of a town or city are those where low-income and marginalized communities live. High levels of pollution can be due to proximity to motorways, fracking sites and industrial manufacturing facilities.

Even though the health studies are there, fossil fuel companies are still knowingly placing facilities like these near marginalized local communities, and will continue to profit at their expense. Not all countries have open fracking sites and this issue varies for everyone across the globe, but the fossil fuels we use come from somewhere, which means they are affecting a local community somewhere.

ACTIONABLE STEPS

- ⊘ Advocate for the end of fossil fuel extraction and other polluting industries.
- ⊘ Understand your town and city and how air quality is possibly worse in some areas than others. Notice what that reflects about the area's demographics or which facilities operate in that area.
- ⊘ Limit vehicle use where possible and take public transport, cycle or walk instead.
- ⊘ Contact your MP or UN ambassador to voice your concern for air quality and ask what air pollution policies they support.

KEVIN J PATEL

Founder and executive director of OneUpAction and climate advocate

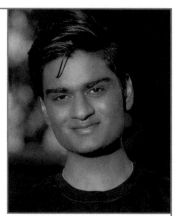

Can you tell us a bit about the Los Angeles community you grew up in and why Black, Indigenous and People of Colour (BIPOC), low-income and marginalized communities are disproportionately affected by climate change?

I grew up in South Central Los Angeles, which is a sacrifice zone, meaning we don't have the resources or the funding to fight against corporations, industries and people in power coming in and putting our communities at risk. Black and Brown communities are literally having to go outside and breathe chemicals and harmful toxins as well as be hit with the air and smoke pollution, heat waves and all of these other factors that, among other things, give our community members underlying health issues. In sixth grade, I was affected by heart palpitations, irregular heartbeat, because of the air pollution that ravages Los Angeles. The reason BIPOC communities are directly impacted by the climate crisis is because we lack the power to stop corporations like the fossil fuel industry from coming into our backyards. You will never see an oil well in an affluent community because, guess what? They have the money, they have the resources, and that's where the politicians usually live.

What is one piece of actionable advice you would give?

A lot of us tend to forget community when we're doing activism; working with our communities, literally seeing what our communities need, the issues that our communities are facing. Any mutual aid that is happening within our communities, like supporting and raising funds for those issues, can really make a huge impact. You can then take it a step further by calling your elected officials and start talking about, 'Hey, these are the issues that we're facing. You need to start implementing these solutions'. Or even coming up with solutions and pitching them yourself. Calling your elected officials, coming up with your own solutions, working with your community, and mutual aid with organizations that are doing amazing work on the ground is where I would begin.

What instills hope in you?

All of the young people who are working intergenerationally and intersectionally, that instills hope in me. We're beginning to understand that we're not the leaders of tomorrow; we're the leaders, activists and solutionists of today. We're really bringing the change that we want to see into reality. We're not waiting around for people to do it; we're doing it now.

NATURE'S SYSTEMS AND HOW THEIR HEALTH AFFECTS OUR HEALTH

We are deeply interwoven with all the varying ecosystems that make up our planet. Their health is our health.

OCEAN HEALTH IS OUR HEALTH

I have a deep love for the ocean; I am humbled by its beauty and force. Ocean health is the topic that brought me into the climate movement, and just like the ocean, this topic is complex, dark and mysterious. So much happens at sea that we don't see – from the unimaginable beauty of the deep-sea floor (a lot of which has never been seen by the human eye) to the heartbreaking slave trade that is happening on some fishing boats. On the research trips I took with Greenpeace in the Gulf of Mexico, the Biscayne Bay in Miami and on the river Wye in England, we trawled for microplastics and found them in every single sample we collected. This made the abstract personal. It confirmed to me that our actions and the health of the planet are inextricably linked.

We often talk about our atmosphere being irreversibly damaged by greenhouse gas emissions, but we talk less about the fact that our oceans are also absorbing these emissions and warming temperatures. According to the Intergovernmental Panel on Climate Change (IPCC), a staggering 93.4 per cent of global warming is absorbed by the ocean as opposed to 2.4 per cent by the atmosphere. This rise in temperature changes the chemistry of the oceans and disrupts marine ecosystems that provide people with food and the oxygen we breathe. At least 50 to 80 per cent of the air we breathe has been produced by plants in our oceans, according to the National Oceanographic and Atmospheric Administration (NOAA), and these same plants sequester carbon dioxide, too. The ocean is both Earth's thermostat and its windows. Without its ability to regulate our climate and feed us with oxygen and nutrients, we would be lost. An unhealthy ocean means unhealthy humans.

The key issues facing our oceans are:

Climate change

➤ The ocean absorbs greenhouse gas emissions, which increase its temperature, disrupting marine life.

- Ocean acidification is happening as the ocean's pH is dropping to the point where life within it is struggling to stay alive.

- Coral reefs are dying due to these changing temperatures, destroying habitats for marine life and degrading the health and biodiversity of the ocean.

Plastic pollution

- Single-use plastics – a mix of consumer products and fishing nets – have ended up in almost every corner of our oceans. Plastic never fully decomposes – it only gets smaller. Plastic harms the marine life that keeps our oceans healthy and thriving.

- Plastic doesn't just get ingested by marine life. As we are all connected, it works its way into human bodies, too, mainly through our food and water. A report by Australia's University of Newcastle suggests that people across the globe consume approximately 5 grams of plastic per week, equivalent to the amount that makes up a credit card.

- The global exportation of waste and recycling to other countries puts stress on their already strained waste management infrastructure. People are forced to live close to open dumps or incineration sites, a threat to their safety and health.

Toxic pollutants

- Agriculture uses chemical fertilizers that run off the land into rivers and then out into the ocean. This creates dead zones, often at the mouths of large rivers, and can destroy entire habitats.

- Toxins and harmful chemicals are flushed through our waterways. Much of this pollution cannot be cleaned out at water treatment plants before heading out to the ocean.

- Waste that ends up in our oceans can leach chemicals into the water and into plants, animals and human bodies.

Fishing industry

- Commercial fishing is depopulating our oceans and practising highly destructive methods, such as bottom trawling, that destroy entire ocean ecosystems.

- ⊗ Overfishing threatens food security for hundreds of millions of people.

- ⊗ Fishing subsidies are estimated at $35 billion a year, of which $20 billion is directly spent on fishing industries that contribute to overfishing, creating economic incentives for the fishing industry to keep exploiting fish populations.

Offshore drilling

- ⊗ The infrastructure required for drilling disrupts marine ecosystems.

- ⊗ Oil spills can be deadly to animals and humans.

- ⊗ The poor air quality and pollution drilling creates can also harm nearby coastal communities.

- ⊗ Emissions from offshore drilling contribute to climate change, and the resulting excess heat is then absorbed by the ocean.

Human rights

- ⊗ As fish populations have been depleted, fishing companies pay their employees less than a living wage, and some workers are forced into unpaid labour, like slaves.

- ⊗ Fishing vessels illegally come into off-limits waters and take fish stock from local communities, who are then forced to hunt inland, altering the balance of the food sources they have relied upon for generations and disturbing the local wildlife. This can also expose humans to animal-borne viruses and other diseases.

- ⊗ Rising sea levels will displace millions of people.

There's a lot of good happening out at sea, however. As I direct my gaze to the ocean's ability to regenerate and look for guidance from marine scientists and innovators, I am inspired and hopeful for our future. If we turn the tide and radically lower our output of greenhouse gas emissions, overfishing and general human disruption to our oceans, they could bounce back and become nourished and thriving landscapes once again. Scientists have said that we need to protect at least 30 per cent of our oceans to avoid the worst effects of climate change. The Global Ocean Treaty proposed by the United Nations would provide this type of protection and chance for ocean regeneration by creating a more resilient network of Marine Protected Areas.

Trawling for microplastics with Greenpeace in the Biscayne Bay in Miami and the river Wye in England. Examining data samples of microplastics aboard the *Arctic Sunrise*.

ACTIONABLE STEPS

- ⊘ Work alongside conservation organizations and your local and central government to support the creation of ocean treaties and sanctuaries.

- ⊘ Vote for candidates who support good ocean policies.

- ⊘ Understand the link between the fossil fuel and petrochemical industries' greed and the plastic epidemic affecting ecosystems and people's daily lives. It did not happen by accident. I highly recommend watching the documentary *The Story of Plastic* and the PBS *Frontline* episode 'Plastic Wars' (season 38, episode 15).

- ⊘ Support the United Nations proposal to end fishing subsidies by writing letters to your UN ambassador and prime minister to stress the importance of this issue. Right now the UK has mentioned making amendments to the subsidies but has not supported the proposal to end them. You can amplify this message by supporting the open letter from the World Trade Organization to stop funding overfishing.

- ⊘ Conserve water. Advocate for cities to allow less runoff and wastewater to flow directly into the ocean.

- ⊘ Reduce the pollutants you use in household products, choose nontoxic chemicals and dispose of chemicals correctly.

- ⊘ Call on your favourite brands to create less single-use plastic waste and transition to refill systems. The pressure to change should not be entirely on us as customers.

- ⊘ What we put in our rubbish and recycling bins never truly goes 'away'. Work to reduce your waste output and learn more about what can and can't be recycled. (More on page 29.)

- ⊘ Eat less fish and more plants. It is challenging to accurately trace if the fish you buy was caught sustainably using nondestructive fishing methods.

SOIL HEALTH: SOIL IS ALIVE, JUST LIKE US!

The health of soil is my second greatest love and avenue of curiosity after the health of our oceans. Soil's vital role in our existence and survival is very similar to that of the ocean: they both sustain life by feeding us and allowing us to breathe, they regulate our climate and they are an incredible solution for the drawdown of the carbon dioxide already in our atmosphere. There are also genuine solutions we can participate in to help these two environments regenerate, but we need to act now.

Since the Industrial Revolution and the conversion of natural ecosystems into land for agriculture, the levels of carbon in our soil have been depleted. But not all agriculture is destructive. If more farms adopt regenerative farming practices that combine Indigenous knowledge, sustainable management and modern research, farmland can be transformed into an environment that promotes biodiversity and rebuilds the health of the soil.

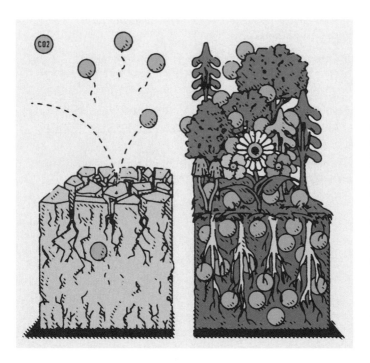

On the left is dry soil completely devoid of plant life. Without plants photosynthesizing, the carbon dioxide stored in the soil escapes into the atmosphere. On the right is healthy soil with lots of plant life. Through photosynthesis, the plants sequester carbon dioxide and store it in the soil.

⊘ Volunteer at a local community garden to learn more about building healthy soil.

⊘ Create your own garden, whether a single pot or an acre.

⊘ Shop at a farmers' market to be connected to local land and the farmers working it.

⊘ Look up local organizations that focus on soil health and regeneration and see how you can get involved. Ask whether these organizations work to strengthen BIPOC land ownership and community.

⊘ Watch *Kiss the Ground* on Netflix, a documentary on regenerative farming.

FOREST HEALTH: THE LUNGS OF OUR LAND

Deforestation refers to the loss of trees, whether human-driven or natural. Forests cover about 30 per cent of the world's land mass, but sadly, they are being destroyed at an alarming rate. Trees are not just beautiful and awe-inspiring to look at; we and so many other species need them. One of the most incredible things trees can do is absorb the carbon dioxide we exhale and the greenhouse gases we create. Trees then create oxygen as a byproduct of photosynthesis. They also provide richly diverse ecosystems and habitats. According to *National Geographic*, 250 million people living in forests and savannas depend on them for nourishment and income, and 80 per cent of Earth's land animals and plants live in forests.

The main industries and human actions that cause deforestation are:

⊙ Agriculture, particularly soybean farms, palm oil plantations and cattle ranching

⊙ Unsustainable forest management

⊙ Illegal and legal logging for tree-based products such as paper and lumber

⊙ Urbanization, as cities expand

⊙ Mining and drilling of metals, minerals and fossil fuels

The effects of deforestation include:

- ⊙ Loss of biodiverse ecosystems
- ⊙ Changes in weather patterns that connect to the overall climate of the planet
- ⊙ Wildfires
- ⊙ The destruction of the temperature-regulating powers of the forest, which affects global water cycles and water supply
- ⊙ More carbon dioxide in the atmosphere, as fewer trees are available to absorb and sequester it (even the process of cutting down a tree adds carbon dioxide to the air)
- ⊙ Loss of millions of people's livelihoods and homes

This all sounds alarming, and it is, but as I witnessed on a trip to the Maya Biosphere Reserve in Guatemala, there are genuine solutions, such as sustainable forest management, that can not only protect forests but also foster community leadership and local economies. The people I met at the reserve had a deep respect and relationship to their land – the work they do as a community is an investment in the survival and happiness of future generations. I think we can learn a lot from how they make decisions with community and family in mind, the rain forest being a member of that family, too.

ACTIONABLE STEPS

- ⊘ Support and amplify Indigenous voices from forest communities who have the knowledge and skills to manage forest biodiversity correctly.
- ⊘ Consider volunteering at or donating to a forest conservation organization.
- ⊘ Look for products certified by the Forest Stewardship Council and Rainforest Alliance, which they consider sustainable to forests and the forest communities.
- ⊘ Avoid products with ingredients from the soy and palm oil industries, which are responsible for deforestation. World Wildlife Fund has a palm oil scorecard resource for consumer brands.
- ⊘ If you eat meat, find out where and how the animal was raised. The cattle industry has deforested large areas to raise cattle. Poor grazing practices degrade the soil, destroying root systems and robbing the earth of its carbon supplies.

MAGDALENA LÓPEZ

Co-ordinator of non-timber products in Uaxactún, a co-operative in the Maya Biosphere Reserve in Guatemala

Why is the concessions model such a great example of sustainable forest management?

We are an example for the world because this model is a conservation model. We are taking advantage of these resources without falling into ambitious extractive activities – we can use them all the time and we can still conserve them. Without these community concessions, it's almost certain that this rain forest would have disappeared already. We also understand that without them, we would not have been able to survive, because we depend on the forest. We have learned to live in the forest and be part of the forest, but also the forest has learned to live with us.

What piece of actionable advice would you give?

I would ask that people don't get tired and continue fighting to conserve the forest. We have to work to protect the very little forest that we still have so we can leave it as a legacy for our children. We are teaching them the importance of conserving this forest. That's why everything has been working so far. Both men and women are committed to that. That's why this also makes us want to improve everything day by day.

What instills hope in you?

What instills hope in me is my family, my children, my community, my people who keep on going, who work for the benefit of all, for the well-being of all, to have a hope in life, to live in harmony with ourselves, among ourselves, and also with the forest.

Touring Miramar Landfill and EDCO's Materials
Recovery Facility in San Diego.

HUMAN SYSTEMS: THE GOOD, THE BAD AND THE UGLY

We have designed many systems that were intended to serve human civilization, but in a lot of cases, their lasting negative effects on both the planet and all living species far outweigh their benefits. Many of these systems we love for their convenience, such as waste management, but how long will this solution feel convenient when we could end up literally burying ourselves in waste?

WASTE RECOVERY SYSTEMS: WE THROW OUR STUFF 'AWAY', BUT WHERE TO?

When I first started to question why rubbish, mainly single-use plastics, was littering our beaches and oceans, my curiosity led me to the waste management system. I have always been drawn to things out of sight and somewhat mysterious (this is why I love the ocean). Weirdly, our waste and where it goes is a little like that: a mystery unless we ask questions. I think it is our responsibility to know what happens to our waste, as we are each likely to throw something in the rubbish or recycling every day. For so long, I had no idea what truly happened with all this stuff. According to the World Bank, the world generates 2.01 billion tons of municipal solid waste each year. So where does it all go?

The answer is both simple and complicated. Let's visualize what happens to our rubbish once we 'throw it away'.

What's a little more complicated is the fact that waste management and recovery is very different across the globe, and some of the waste and recycling created in places like the UK is shipped to other countries. According to the World Bank, 90 per cent of all waste is openly dumped and burned in low-income countries, and their poor and vulnerable communities are disproportionately affected. The waste can also run off the land and into waterways that head out to the ocean. This is called nonpoint source (NPS) pollution and is one of the main ways rubbish and plastics end up in our oceans. These open waste dumps can quite literally bury homes and people when the infrastructure can't hold them, and they have even been known to explode due to the build up of hazardous and toxic waste.

ACTIONABLE STEPS

- ⊘ Consume less and waste less.

- ⊘ Fix and mend broken items rather than throwing them away.

- ⊘ Know the materials your household items and packaging are made from and the environmental implications of those materials.

- ⊘ Compost organic waste, and advocate for curbside compost programs.

- ⊘ Put pressure on your local government to make sure waste facilities are not located close to residential communities.

- ⊘ Since 2018 the UK government has been promising the implementation of a bottle deposit scheme to reduce marine plastic waste. Lobby your local MP to raise this as an urgent parliamentary issue.

WASTEWATER SYSTEMS: DON'T FLUSH THAT!

Again, this is a system most of us don't really know much about. We conveniently flush and don't think twice. But so much happens after that liquid waste leaves our homes! The water (and anything in it) from our toilets, sinks, dishwashers, washing machines and showers goes to a water treatment plant before being sent back into our waterways and oceans. I toured the Hyperion Water Reclamation Plant, which sits right on the Pacific in Los Angeles and receives 275 million gallons of wastewater on a dry-weather day. Hyperion has some pretty innovative methods (including big oval combustors that convert the methane gas from our poo into power for the plant!), but sadly, not all toxins can be treated to make this water totally clean and drinkable again. It was eye-opening to see all the things people put down their drains! In order to preserve water, we need to remember that without this precious resource, life as we know it would not exist, and it is our job to help keep our waterways clean and healthy.

ACTIONABLE STEPS

- ⊘ Only flush something down the toilet if it has passed through you. That means no tampons, nail clippings, contact lenses or paper products other than toilet paper.

- ⊘ Wash your synthetic clothes using products that capture the microplastics they shed to stop them from entering our oceans. (More on this on page 70.)

- ⊘ Use nontoxic and ecofriendly detergents and soaps.

- ⊘ Read your labels. If you think a household product could be hazardous, it probably is. Check with your local council to see when your hazardous waste pick-up is or if you have to take it to a local drop-off site.

FOOD SUPPLY CHAIN:
A SYSTEM WE ARE DEEPLY WOVEN INTO

The food supply chain describes how food from a farm, sea or river ends up on our table and the systems it travels through to get there. This can include production, processing, distribution, consumption and disposal. Our food system is governed by supply and demand, so as demand fluctuates, the whole system can be impacted.

There are many things to consider when weighing up our food choices. Every choice we make as consumers affects the farmers growing the food, the environment in which the food is grown, the communities surrounding those farms and the transportation companies moving the goods. And these are all significantly affected by our changing climate. We each choose the foods we eat for various reasons, including our access to food, our values, our beliefs, our taste buds. I think it is important that we understand and take responsibility in knowing how far that food has travelled to get to us and the impact of the agricultural methods used to produce it.

We have developed a codependent relationship with the food supply chain. The accessibility of different types of foods from all parts of the world, at any time of year, has made us less resourceful and self-sufficient. The more urbanized we have become, the less we have grown our own food or developed direct relationships with farmers, and the longer and more complex our food supply chain has become. This disconnect has also led to food apartheid in many lower-income communities, where larger chain supermarkets and fast-food restaurants have pushed out local markets and businesses, making nutritious and affordable food less accessible.

A shorter supply chain would have the following positive effects on both the planet and community:

> It would lower carbon emissions due to less long-distance transportation.

> Less packaging is needed when a farmer sells directly to a customer.

> Growing your own food can liberate you from having to pay for it.

> It creates community empowerment and better relationships within the community.

> It results in greater education on what our food is and where it comes from, and respect for those who work in agriculture.

> Farmers can earn a better living.

How to shorten your food supply chain. Working from the most extreme (the outer red ring), which could be both buying produce from a foreign country and having it delivered to your door, to buying produce grown in your own country at a supermarket, to buying locally grown produce at a farmers' market, to growing your own food at home.

- ⊘ Grow some of your own food, even if it's one herb plant in a pot.

- ⊘ Sign up for a local community supported agriculture (CSA) share, or a vegetable box delivery scheme, and shop for produce at a local farmers' market. This helps support local agriculture and reduces transport emissions.

- ⊘ Choose to shop at individually owned greengrocers and local food shops rather than at the larger supermarket chains.

- ⊘ Support BIPOC farmers and food establishments to diversify the food system and invest in the community.

- ⊘ Support the National Food Strategy's call for a sustainable, healthy food system in the UK, including ensuring all local authorities in England develop local food strategies.

- ⊘ If you're shopping online for food, try to do one large order rather than multiple smaller orders.

- ⊘ Cook more meals from scratch with whole ingredients instead of relying on premade packaged and processed foods.

- ⊘ If you are ordering a takeaway from nearby, walk to the restaurant and pick it up yourself.

- ⊘ Follow a diet of moderation and balance out your food groups.

The Ecology Center

For a lot of us living in urbanized areas, the farms where our food is grown are usually out of sight and out of mind. However, as we begin to understand how broken our food supply chain is and how beneficial growing food can be for communities, I am seeing more and more inner-city farming and gardening projects pop up across the world. One of my favourites to visit is The Ecology Center, a 28-acre regenerative farm in Orange County, California, founded by Evan Marks. I love to shop at their incredible farm stand, located directly next to the farm, so I can see where all the beautiful produce was grown. They also offer CSA boxes and programme a wide variety of educational events, which include in-depth farm tours, lessons on composting and growing your own food, an elevated u-pick experience for children and community dinners. Plus, they donate 20 per cent of their produce to neighbours in need. In their ten-year history, 180,000 people have engaged with the farm, which to me is an amazing number of people to experience firsthand how food grows and our relationship to it.

Visiting The Ecology Center with founder Evan Marks during their summer TASTE Festival.

SOCIETY AND ECONOMIC SYSTEMS

We don't choose where we are born or which society or economy we are born into, yet these radically alter our experience of the climate crisis.

CLIMATE JUSTICE: THE INTERSECTIONAL APPROACH WE MUST ADOPT

For a long time, the narrative around climate change has been 'Save the planet', but what about the people? Calling climate change a climate *crisis* more accurately depicts the fact that people's lives are being threatened and our need to act is urgent. We need to recognize that those currently at risk are disproportionately affected due to racial and socioeconomic factors. According to the environmental and climate justice programme page on the NAACP's website, 'Race is the number one indicator for the placement of toxic facilities in this country'.

The effects of climate change are disproportionate not only within the United States, but from one continent to another. The global north has created more carbon emissions than the global south, yet the south is being affected by climate change first and most extremely. Until we recognize our global connectivity and respond collectively, we will not harness our potential as a species to address the climate crisis and our future on this planet. This is why an intersectional approach to the climate crisis is essential. Taking note from platforms such as Intersectional Environmentalist and People & Planet, an intersectional approach to the climate crisis would call for:

- Creating safe and healthy environments for all communities, regardless of race, income, sex or ethnicity.

- Dismantling all systems of oppression – all systems overlap and therefore impact vulnerable communities.

- Recognizing the Black, Indigenous and People of Colour (BIPOC) communities that have been bearing the brunt of the climate crisis and centring them in the climate justice movement.

- Acknowledging that every community is different and has different needs; each needs to be respected and action designed with those needs in mind.

The environment and conservation movement has predominantly been led by White and privileged people. I recognize that I am part of this. In recognizing this, I pledge to actively learn about environmental racism, share what I learn with my community and use my privilege to advocate and amplify the voices of Black, Indigenous, People of Colour and their communities.

ACTIONABLE STEPS

- Support and encourage representation in organizations, governments and corporations.

- Work to centre BIPOC voices in the movement, at boardroom tables and in government cabinets.

- Amplify and listen to BIPOC climate activists and communities, both local and international, who are already experiencing the damaging effects of the climate crisis.

- Notice where you may have experienced privilege in climate-related issues, such as access to education on the subject, clean water, quality food and green spaces.

- Look to organizations such as Intersectional Environmentalist as a place for resources, education and community.

- Encourage businesses, governments and committees to adopt assessments such as the Racial Equity Impact Assessment (REIA), a systematic examination of how different racial and ethnic groups will likely be affected by a proposed action. (More at raceforward.org.)

LEAH THOMAS

Climate justice advocate and founder of
Intersectional Environmentalist

Since you started Intersectional Environmentalist, have you seen an evolution or change in the environmental space, in the awakening in education around the issues of intersectionality and climate justice?

I couldn't have imagined from last year into now the conversations that are happening at some of the world's largest environmental organizations. Some that are primarily focused on conservation are even reaching out to us and saying, 'Oh, we realized that we also need to focus on the people component'. I've also seen curriculums change at universities, like the addition of environmental justice classes. I've seen a lot of change, but I've also seen the 'same old, same old' people who are very set in their ways of what environmentalism needs to look like.

What is one piece of actionable advice you would give?

I would say attitude-wise, progress over perfection. When I started my environmental journey, when I made that switch from talking about skincare to climate justice to activism to going into protests, those first couple of months, I definitely had a very 'this is right, this is wrong, do this, do that' approach. It's easy to get sucked up into that with activism. I would say leaving room for that nuance, having that intersectional approach, and remembering that people might be beginning their sustainability journey and we're constantly learning. Get curious, have fun and know that you're going to be wrong and it's totally okay as long as you just evolve in the process.

What instills hope in you?

The younger generation gives me a lot of hope. They just *get* intersectionality so much more because they're already having conversations about race, religion, gender, sexuality. Applying that to their environmentalism is an invigorating way for them to say, 'Oh, I can be an environmentalist and talk about all these different topics I like'. That's really cool.

CLIMATE MIGRATION: AN INCREASING REALITY FOR MILLIONS OF PEOPLE ACROSS THE GLOBE

One of the biggest threats to the human species is being displaced from our homes due to the devastating effects of the climate crisis. That could be large-scale destruction due to extreme weather events, the loss of drinking water, food insecurity, drought or unemployment. This type of displacement is already happening at a significant scale, yet we do not hear enough about it either because the most influential cities are not yet feeling the effects or because capitalist nations are not eager to admit that the polluting industries they support and rely on are the main cause.

The United Nations reports that there are roughly 26 million climate refugees – and estimates that more people are displaced due to climate-related issues than war. The big question with all migration issues is, Where will people go? A lot of migrants will be moved internally within their home country, which puts pressure on other areas and their resources. If migrants have to leave their country, which countries will allow them to seek asylum?

ACTIONABLE STEPS

⊘ Educate yourself on the issue and recognize that this is happening. Avoid shifting blame from government to government, as this could perpetuate xenophobic and outdated colonial power relations. This can distract us from what really matters: helping vulnerable people.

⊘ Call your representatives and encourage them to support policies that protect migrant rights and stop polluting industries from creating the conditions and environments that are forcing people to flee in the first place.

⊘ Volunteer with local nonprofits helping people settle into their new country of asylum.

⊘ Donate to organizations such as Choose Love. Every year my gifts to friends and family over the holidays are from Choose Love. Gifts range from $7 to $120, and you can choose specific things to be given directly, such as warm blankets or mental health care.

INDIGENOUS KNOWLEDGE AND LIBERATION

I will admit that until living in the United States, an occupied country that was colonized and taken from the Indigenous nations of this land, I had never truly and deeply thought about how oppression of Indigenous peoples and suppression of their cultures and relationship to the land can cause climate devastation. Indigenous peoples comprise less than 5 per cent of the world's population yet protect 80 per cent of Earth's remaining biodiversity, making them some of the most powerful and essential protectors of the last biodiverse frontiers. As we face the climate crisis, we are beginning to realize that Indigenous knowledge is central to our future relationship to Earth.

Inspired by the paper 'Kincentric Ecology: Indigenous Perceptions of the Human-Nature Relationship' by Enrique Salmón, anthropologist and head of the American Indian Studies Program at California State University, here are some of the perspectives we can adopt and support in an effort to learn from and honour Indigenous knowledge:

Nature as kin, not other – placing ourselves within nature, as part of it and not separate from it. Treating Earth like we would treat a relative.

Biodiversity – fostering biodiverse ecosystems to increase resiliency.

Agroecology and agroforestry – an integrated approach to agriculture and forest management combines science, knowledge and tradition. Its principles include diversity, co-creation, sharing of knowledge, resiliency, circular solidarity and economy.

Sovereignty – the right for Indigenous peoples to govern themselves and practise their own systems of leadership and belief.

Intergenerational wisdom and knowledge – the knowledge and experience of the land, which has been developed over generations, is traditionally passed down directly through storytelling, skill-sharing, language and experience within the context of the environment. It is important that we call on this firsthand experience.

ACTIONABLE STEPS

- ⊘ Learn more about Indigenous nations' land, either where you live or where you are travelling to. Native-land.ca is a great resource for this; you can search an address and explore the interactive map.

- ⊘ Tell the true stories of Indigenous erasure, and advocate for its depiction in history books.

- ⊘ Centre Indigenous voices in the climate movement and across all related industries such as agriculture, forest management and policy.

- ⊘ Take responsibility to be a steward of the planet, and be involved in its care and upkeep.

- ⊘ Show up in person to experience knowledge and wisdom rather than limit yourself to the internet, media and books.

- ⊘ Support representation of Indigenous peoples in government, organizations and committees.

- ⊘ Learn more about the Landback movement, and consider options you think you could make reparations for, such as returning land.

go see

OBSERVING THE HABITS YOU PRACTISE AT
HOME CONNECT TO THE ISSUES AT LARGE

Once we have familiarized ourselves with the big picture of the climate crisis, it's time to transition to the four walls of our home. I am super-excited to invite you into my home and share my experience of working to be gentler on the planet and myself. This is a continual and never-ending journey, so we are learning together. The practices I have been implementing in my home have been a private and solitary act until now. I was called to write this book because I wanted to share these quiet acts that brought joy and a sense of connection to the wider issues. Over time, I have been learning and gathering tips and practices from books, blogs, videos and social media. No matter what we do for a living or for fun, each of us spends a significant amount of time in our home. It is liberating when this important space begins to reflect and represent our values – what we truly believe.

Let's start by walking around and looking at the various spaces in our homes. Taking on your entire home all at once can be overwhelming and lead to feeling out of control, a feeling I hope this book changes for the better. My advice would be to tackle one area of your home or one issue first. In my journey, I started by looking at packaging within my kitchen as I became increasingly aware of the damaging effects single-use plastics have on ocean health. Once I began making small changes in my kitchen and seeing the positive impact on my rubbish bin and mood, I was hooked on researching and using my creativity to find more positive habits I could adopt. This excitement led me to look at single-use packaging in my bathroom and then in my house as a whole. My latest fascination is compost and soil health. For you it could be something entirely different, such as the fibres in your clothing and where and how it was produced.

The most helpful tool for me was an inventory table. I started by writing down five to ten items in a given area of my home, then evaluating them based on their materials, functionality and environmental impact. This helped me to figure out which of my habits I wanted to evolve and how to implement those changes.

I want to stress that this journey is not about getting rid of everything and starting again in order to adopt a perfect 'sustainable lifestyle'. Instead, I want to encourage you to be creative with what you already have and to find methods and tools to get the most out of what you have, buy, make and do. There is no right or wrong, only suggestions, facts and options to be inspired by. Throughout this book, we will not only shift habits to reduce our carbon footprint but also form habits to maximize our use of resources.

As I started with my kitchen, let's begin there.

THE KITCHEN

The kitchen is the brain of the house: lots of decisions are made here, and it's where we find our nourishment. But sadly, a lot can be thrown away here, too. However, once I started thinking critically about what food I buy and how I buy it, I soon discovered that many positive changes can be made here. Today I am getting the most out of the food I buy, buying food that supports farmers and the land, composting my food waste, storing and keeping my food in clever ways, and much more.

We are all creatures of habit. There will be some items you are attached to buying, but there is also opportunity to find alternatives, discover new things to become attached to, use what you already have and sometimes compromise. I invite you to get out a notepad, a scrap piece of paper or a note-taking app to make notes as you take an inventory of your kitchen and begin to get to know your space, habits and potential for solution-driven change.

BUYING HABITS: HOW YOU SHOP FOR FOOD

I started by identifying which food and kitchen items I bought frequently. According to the United Nations, a shocking 1.3 billion tons of food that is produced for human consumption is lost or wasted globally, which is one-third of the food produced. That is an awful lot of sun, water, energy, labour, transport emissions and nutrients wasted. Why does this happen? It's a combination of the rigid food supply chain, overbuying by supermarkets and consumers, people storing food incorrectly and the repercussions of this waste (from wasted finances to the methane emitted by landfill).

When identifying my food-buying habits, these were the questions I asked myself. Here is the first inventory table I ever did:

BUYING HABITS INVENTORY TABLE

ITEM	MATERIAL & PACKAGING	WHERE & HOW IT WAS GROWN/ PRODUCED	HOW OFTEN DO I BUY IT?	ENVIRONMEN-TAL IMPACT	OPINION	SOLUTION
Honey	Plastic jar and lid	UK	Every 4 weeks	Packaging made from virgin plastic, which means fossil fuels extraction.	Not ideal, worry the plastic isn't recycled.	Find a nonplastic alternative.
Coffee	Mixed-material plastic and paper	Central America, organic	Every 2 weeks	Packaging unrecyclable. Living conditions of farmer?	Researched company and found good labour rights details.	Buy larger bulk-size bag or find a refill option selling loose beans. Compost grounds.
Olive oil	Glass bottle and metal lid	Greece	Every 3 months	Recyclable. Sometimes has plastic seal.	Mostly positive.	Buy as local as possible and without plastic seal.
Eggs	Cardboard	UK, free-range	Every 2–3 weeks	Package recyclable. What is the best type of egg?	Mostly positive, I think?	Look for pasture-raised and organic.
Cheese	Plastic	France	Weekly	Plastic won't decompose in landfill.	Negative.	Look for plastic-free alternative from a local shop. Source local cheese.
Cherry tomatoes	Plastic	Organic	Every other week	Can this type of plastic be recycled?	Can easily improve.	Buy locally, so only when in season. Find loose tomatoes not in packaging. Or grow my own?

This simple exercise helped me ascertain which issues mattered most (for me, at first it was cutting down on single-use plastic) and where my shopping habits could be improved. Over time, I have shifted my habits, working towards the solutions column every time I'm at the supermarket.

STORING FOOD TO MAXIMIZE ITS POTENTIAL

I think one of the easiest tricks to get the most out of what you buy and not be wasteful in the kitchen is to store ingredients correctly and more efficiently. I have loved learning clever tricks over the years to get a longer life out of my food and now have a much more organized kitchen. For instance, I've figured out how to make those big bunches of herbs last for a week or more (see page 180), a major improvement from the usual two-day shelf life I was getting out of them before. I've also figured out how best to use my freezer to my advantage and how to keep my pantry organized. I have learned which produce is happiest stored at room temperature on the worktop and which prefers a cool, shaded cabinet or cupboard. A great way to start is by using an inventory table like the one we used for buying habits. In this example, we're looking at how the item was packaged when we bought it, how we're currently storing it, how it's looking, its impact on the environment and possible solutions. Here's the first one I did:

FOOD STORAGE INVENTORY TABLE

ITEM	ORIGINAL PACKAGING	WHERE/HOW IS IT BEING STORED?	HOW IS IT LOOKING?	ENVIRONMENTAL IMPACT	POSSIBLE SOLUTIONS
Half an avocado	Bought loose at the supermarket and put directly in my basket	Wrapped in cling film in the fridge	A little brown	Negative, as the plastic will never decompose in a landfill.	Store in a reusable silicone bag or upcycled container.
Potatoes	Put these in a plastic produce bag at the store	In the plastic bag in the fridge	Soft, don't look great	Didn't need the plastic bag.	Buy loose, without plastic bag, and store in a cool, dark cupboard.
Veggie burgers	Individual plastic pouch and cardboard box	In original packaging in the freezer	OK	The cardboard is recyclable, but the plastic is not.	Make my own and store in a reusable container.
Almonds	Resealable plastic bag	In an upcycled jar in the pantry	Good	Plastic will not be recycled or decompose.	Seek a plastic-free bulk-buying option or at least buy in larger quantities. Choose a less water-intensive nut?
Beans	Steel tin	In original tin in the pantry	Good	Positive – steel can be recycled when tin is clean. Long shelf life.	Make sure to clean the tin before recycling.
Bread	Soft plastic bag	In original bag on the worktop	A little stale and sweaty	Plastic will not be recycled or decompose.	Bake my own, or buy in brown paper bag at a bakery or farmers' market. Store in a more airtight container.

Creating an inventory helped me think through my storage solutions for every aspect of my kitchen. By experimenting with different products and what I already have in my kitchen, I have developed the following tips and actionable steps to take.

Fridge:

- ⊘ Refrigerating some fruits and vegetables – such as tomatoes, strawberries and aubergine – too early can kill their flavour and stop them from properly ripening. Leave them on the worktop instead.

- ⊘ Some vegetables, such as potatoes and garlic, and fruits such as bananas, never need to be refrigerated and instead do better either on the worktop or in a dark cupboard.

- ⊘ Skip the plastic produce bags at the supermarket; instead, bring your own reusable cloth ones, or put the loose produce in your basket.

- ⊘ When storing leftover food, use reusable silicone bags in place of disposable plastic ones, reusable wax wrap instead of cling film and upcycled jars or reusable glass, metal or bamboo storage containers.

Freezer:

- ⊘ Many premade frozen meals are heavily packaged in plastic. If you're able, try making your own favourite freezer items such as waffles or veggie burgers from scratch and storing them in reusable storage options such as zip-top bags made from silicone.

- ⊘ Instead of buying frozen steamed vegetables, steam your own and freeze in ice cube trays or reusable bags and containers.

- ⊘ If you have fruits that are on the verge of being overripe, freeze them to use in smoothies or baked goods.

- ⊘ A lot of items can get lost in the back of our freezers. It's important to routinely clean out this space to avoid wasting energy freezing food that is long past its edible life. Frozen fruits can be stored for about a year and frozen vegetables closer to 18 months.

Pantry:

⊚ Try to buy pantry items from the loose/bulk section in the supermarket, if available, as this will help eliminate unnecessary plastic packaging. Or buy larger bulk-size portions of your favourite or staple items so there is a higher product-to-packaging ratio. This can be more economical, too.

⊚ Don't buy a whole new set of containers to organize your pantry! Use upcycled glass containers such as old wine bottles and nut butter or jam jars and even old takeaway containers. It often takes a little scrubbing to remove labels and get the jars clean, but I like the odd assortment of jars and containers I've collected over time. (See my jar tips on page 185.)

⊚ Be sure to store ingredients that are sensitive to moisture, such as sugar, in jars or containers with a good seal, to keep them fresh.

Worktop:

⊙ Keep your fruit and vegetable bowls out of direct sunlight.

⊙ Store bread, potatoes, sweet potatoes and alliums such as onions and garlic in a cool, dark place. Keep onions and potatoes separate for the longest shelf life.

⊙ If you store items in fruit bowls, learn which ones are higher in ethylene (the fruit-ripening hormone) or are ethylene-sensitive. For instance, if an apple is stored next to potatoes, the potatoes will ripen quicker and therefore not last as long. (More on this on page 183.)

APPLIANCES, COOKWARE AND UTENSILS

The next area I explored was my collection of small appliances, cookware and utensils. The materials in these each have their own positive and negative effects on our health and the planet. Again, I am not suggesting that you throw away all your cookware and utensils and buy new ones, as that would be wasteful. Instead, I'm suggesting a few things to consider the next time you need an item. If this section inspires you to spring clean and cull your kitchen paraphernalia to just your most-loved items, that's great! Just be sure to research how best to recycle, donate or resell the items you've decided to let go of.

To get a sense of the types of appliances, cookware and utensils I had in my kitchen and their respective environmental impacts, I did another inventory table:

APPLIANCES, COOKWARE & UTENSILS INVENTORY TABLE

ITEM	MATERIAL	PROS	CONS	SOLUTION
Nonstick pan	Teflon coating	Affordable. Mess-free.	PTFE coatings emit toxic fumes that have been linked to health issues such as breast cancer.	Opt for a nontoxic option such as stainless steel, cast iron or ceramic.
Plate	Ceramic	Bought secondhand. Love it, as it's unique!	Make sure there aren't too many cracks or chips.	Great, as I'm not buying new things. Take care of it.
Spatula	Silicone	Helps me get every last bit of food out of my jars!	When washed at a high heat, can shed microplastics, which then enter our waterways.	Wash by hand instead of putting in the dishwasher.
Chopping board	Plastic	Easy to clean.	Can shed microplastics when washed. Bacteria can grow in the knife cuts over time, so needs to be replaced often.	For my next chopping board, opt for wood instead, which can be sanded down or treated to last a lifetime.
Spoon	Wood	Durable. Affordable. Antibacterial properties. Biodegradable.	Wood could have been harvested under poor, unsustainable forest management.	Keep using and caring for my wooden spoons by washing by hand and never soaking. Seek wood certified by the Forest Stewardship Council (FSC).

Once you've completed this inventory table exercise, you can begin to think through positive changes you can make in your kitchen. After some practice and reflection, here are my main takeaways:

- When buying cookware, consider not only which materials were used to make the item and how it performs in the kitchen but also the conditions under which it was made and its afterlife. How can you best care for it and eventually dispose of it?

- Instead of buying new cookware, can you find something secondhand?

- Figure out how to properly care for your kitchen tools and appliances to help preserve both them and our planet. For example, cast-iron pans need special tending, wood can be sanded down and plastic is better washed by hand rather than exposed to the high heat of a dishwasher.

- Do some research to see if there are any groups in your area that get together to fix small appliances such as toasters and food processors. This could also be a great way to meet some other like-minded people in your neighbourhood. If you love this subject, you will enjoy a BBC show called *The Repair Shop*.

- Where possible, choose wood over plastic. It has antibacterial properties, is biodegradable and doesn't shed microplastics when washed.

- Unplug all small kitchen appliances when they are not in use, including coffee grinders and toasters, to save energy.

RUBBISH: HOW DO WE CREATE IT, AND WHAT DO WE DO WITH IT?

We tend to use waste as an all-encompassing term for anything we no longer deem useful, whether that's our kitchen rubbish, a broken appliance or a shirt we no longer wear. And while we'll always use more technical terms such as 'global waste' and 'waste recovery', the most powerful lesson I have learned during this journey is to re-evaluate and work to eliminate the word 'waste' from my everyday vocabulary. By calling items, materials and precious resources 'waste', we strip them of their value and quickly divorce ourselves from having any responsibility to them. After trying to eliminate single-use plastics from my home, I realized I had developed quite an entitled relationship with my rubbish. I consumed what I liked, threw the rest away and washed my hands of it. The convenience of my local council picking it up and taking it away every week worked well for me.

This changed when I noticed not only the pollution in our oceans and how that connects to the overall health of our planet but also how our soils were becoming polluted, too, and losing the nutrients needed to grow food.

RICHARD ANTHONY
President of Zero Waste International Alliance

How do we change our relationship to 'waste' and look at it more as a resource?

The first thing is to stop using 'waste' as a word. 'Waste' is not a noun, it's a verb. I waste, you waste. They waste. It should be 'zero wasting'. Because when you look at so called waste, what is wasted? Well, it's paper, it's not rubbish. But that paper is a resource, it's organic. People understand waste as something that they can throw away. If you don't call it waste – you call it paper, you call it metal – that's a whole different story. A lot of it's how you frame it, for sure.

What is one piece of actionable advice you would give?

My recommendation is to become active, at the very least to take care of yourself. Try to make sure that things you purchase can be recycled or repaired. Once you figure out yourself, then you work on your family, then on your local government. Let your local council member and MPs know how you feel about environmental issues, especially resource recovery and recycling. Get like-minded people together to make democracy work in your own area. The key is that you have to ask. No politicians ever came up with anything by themselves. They've got to be asked.

What instills hope in you?

Well, when I look forward, I see a lot of fear and trepidation in terms of the future. But when I look backwards over the last fifty years, I see a lot of amazing accomplishments, too. Going to a restaurant, a taco shop, and seeing a recycling bin, it shows me the guy realizes it's a resource there. That kind of stuff gives me hope. I'm a glass-half-full kind of guy anyhow. If I didn't have hope, I'm not sure I'd want to be around.

The best way to take stock of what you're throwing away and see how you can begin to be more resourceful and less wasteful is to do a rubbish audit inventory table.

I encourage you to do a similar audit of the rubbish and rubbish systems in your home. Here are some general ideas and guidelines to keep in mind:

⊘ Research which types of materials can be recycled in your area, and opt for those when shopping. Some curbside programmes accept only one or two types of plastic. Check out page 100 for a more in-depth look at these materials.

⊘ If your curbside program doesn't accept a lot of materials, see if there is a local recycling drop-off centre that will.

⊘ If your recycling is collected from your house or apartment building, you can put the recycling directly into that bin rather than use a plastic rubbish bag. Or, opt for a paper bag you already have.

⊘ What are your rubbish bags made out of? Commonly, they are either made of plastic, recycled plastic, bio-based plastic or a mix of these. None of these bags ever truly decomposes in landfill due to their chemical structure, so there isn't a clear best option here. I like to use either bags made from 100 per cent recycled materials or bio-based bags. (To learn more about bio-based plastics, see page 101.)

⊘ Start composting your food scraps! You will be shocked by how much more slowly your rubbish bin fills up once you do. (More on this on page 222.)

⊘ I have personally found that by having a small landfill rubbish bin, I am less likely to throw things out because I can easily see how much rubbish I have created, instead of it being a bottomless pit!

I hope that by cultivating a more informed and thoughtful relationship to the items you use – recognizing the fact that they are resources before they become rubbish – you will begin to see the contents of your rubbish bin decreasing. I have found no greater joy and sense of achievement than throwing away less, less often.

RUBBISH AUDIT INVENTORY TABLE

ITEM	WHERE DO I PUT IT?	ENVIRONMENTAL IMPACT	POSSIBLE SOLUTION
Aluminium tin	Recycling	OK	Be sure to empty and rinse and place the lid (if detached) inside the tin.
Glass jar with metal lid	Separate lid and recycle	OK	Empty and wash. Remove the lid and recycle.
Plastic takeaway container	Landfill bin, as it is #5 plastic	Negative, chemicals in the plastic can leach into the soil	Make my own dinner or eat out at a dine-in restaurant.
Onion peel	Rubbish, because I do not have a compost system	Negative, goes to landfill	Set up my own home compost system or find somewhere to drop it off.
Old magazine	Recycling	OK	Consider upcycling into wrapping paper or donating to a library for a magazine exchange.
Hard plastic tray/lid with fruit such as apples and pears	Recycling, my curbside programme accepts this plastic	OK, but will it ever actually be recycled?	Buy loose fruit instead and bring my own reusable produce bag.
Biscuit packet	Rubbish because it is #7 mixed-material packaging	Negative, goes to landfill and chemicals in packaging can leach	Make my own biscuits or buy ones packaged in recyclable materials.

BEDROOM

There are lots of larger, higher-investment items in the bedroom such as the bed, mattress, linens, furniture, etc., but as we focus on the more everyday items, the most obvious is our clothes. The fashion industry is one of the most polluting industries on the planet and is responsible for 10 per cent of humanity's carbon emissions. That's more than all international flights and maritime shipping combined. According to the International Union for Conservation of Nature (IUCN), 500,000 tons of microfibres from synthetic clothing enter the ocean each year when we wash our clothes. That's equivalent to 50 billion plastic bottles. The damaging effects of the fashion industry are becoming increasingly known around the world, and people are demanding transparency from the brands they buy.

It is becoming fashionable and celebrated to buy and sell secondhand, seek out textile recycling options and swap clothes with friends or family. I have swapped (or should I say 'long-term borrowed'?) items of clothing from my parents, such as my dad Gary's knitted jumpers he's had since he was my age. Currently, the resale market is valued at about $28 billion and is forecast to reach $64 billion in the next five years, according to thredUP and GlobalData. Let's pull back to a scale we can comprehend and see how we can access and participate in a lower-impact fashion industry.

It wouldn't be beneficial to throw all our clothes out and start again 'sustainably' because that would create a large amount of wasted textiles. Instead, let's take a closer look at five items you might find in your wardrobe and their potential environmental impact:

CLOTHING INVENTORY TABLE

ITEM	WHAT MATERIALS? IF FARMED, HOW?	WHERE WAS IT MADE? UNDER WHAT CONDITIONS?	GARMENT-MAKING PROCESSES	SMALL BUSINESS, FAST FASHION, BIG BRAND	HOW DO YOU WASH/CARE FOR IT?	SOLUTIONS
Yoga leggings	Nylon	USA	Material made from recycled plastic bottles. Small-scale factory.	Small business	Machine, microfibres will shed.	Research washing machine product/tool that captures microplastics. Look for natural-fibre alternative.
Wool jumper	Alpaca wool	Peru, Fair Trade Certified practices	Small-scale production with local artisans	Small business	Hand-wash, dry in sun.	Take care of it and it will last.
Jeans	Cotton (not organic), dyed	Italy, member of Fair Wear Foundation	New material, dyeing of jeans is harmful to waterways	Small business	Machine wash, dry in sun.	Mend if they get worn/torn, air-dry them.
Underwear	Nylon	China, unknown working conditions	Most likely harmful	Fast fashion	Machine wash, microplastics will shed. Only lasts 2–3 years.	Research washing machine product/tool that captures microplastics. Natural-fibre and small-business alternatives.
Jacket	Nylon, feather down, plastic zipper	China, unknown working conditions	Likely mass-scale factory. Were the feathers humanely harvested?	Big brand	Have never washed; air it out after camping trips.	Hard to find natural-fibre alternatives. Care for it well. Could have bought secondhand.

This simple exercise is a helpful way to remind yourself of just how many people and materials are involved in making one item of clothing. Here are some things to consider when you're looking through your wardrobe:

> Where is the clothing made and under what conditions? Were the workers paid fairly/provided with housing?

> How far did the clothes travel to get to you?

> What is the impact of the dyeing process used?

> What materials are your clothes made of? How much water was used to grow and/or produce them?

> Is it synthetic or a natural fibre? If natural, was the source material regeneratively grown?

> How can you best take care of your clothing, mend items when needed and, if necessary, donate, resell or recycle them?

When you buy something, try to ensure that at least two of these specifications apply:

> Secondhand/thrifted

> Not from a fast-fashion label (inexpensive clothing produced rapidly by mass-market retailers in response to the latest trends)

> Made locally

> Small business

> Natural fibre, such as wool, cotton, hemp

> Recycled materials

How you care for your items and how you wash them can make a huge impact, whether you fix them when they are damaged or torn, hand-wash them versus dry cleaning them, use tools to collect microplastics and prevent them from entering waterways, or wash clothes on a cold cycle and air-dry them when possible. According to the nonprofit Waste & Resources Action Programme (WRAP), by extending an item of clothing's lifespan by just nine months, you can reduce the waste and water carbon impact of that item by 20 to 30 per cent. (More tips on this in the 'Go Keep' chapter, page 185.)

BATHROOM

Similar to the kitchen, the bathroom is an environment in which we use a lot of products, many of which are highly packaged and can't always be recycled. I found that by doing a mix of making some of my own products, shifting to buying products packaged in recyclable materials and refilling containers, I could seriously limit what I was sending to landfill. Making these changes also directly

and positively affected my health and well-being. It forced me to take some time to slow down through self-care and opt for products with cleaner, simpler ingredients that could be absorbed happily by my skin, which is our largest organ.

When I started to lower my use of single-use plastic packaging in my bathroom, I was overwhelmed with all the products that I owned but never used, had used just once or had small trial packages of. I had the urge to throw everything away and start from scratch, but that would mean a very large rubbish bag going to landfill. Instead, I tried to use up opened bottles of things and asked if friends would like some of the products that I found didn't work for me. After doing a big bathroom-product edit, I did another inventory table:

BATHROOM INVENTORY TABLE

ITEM	MATERIAL & PACKAGING	IMPACT	OPINION	SOLUTION
Toothbrush	Plastic	Never decomposes in landfill. Impact on ocean and land health.	Negative – want to change.	Bamboo toothbrush.
Hand soap	Plastic bottle	Plastic pump is hard to recycle, will go to landfill; bottle might be recyclable.	Negative – easy to change!	Buy loose soap bars or ones packaged in paper.
Face wash	Glass and plastic	Glass is recyclable, pump is not. Ends up in landfill.	Partly negative.	Find refill option or recycle pump with TerraCycle.
Shampoo	Plastic	Will it really be recycled? Ocean and land health.	Negative – want to change.	Shampoo bar or refill.
Razor	Plastic, varying types, and metal	Not recyclable, will not decompose in landfill. Ocean and land health.	Negative – the razor head is frequently replaced.	Seek plastic-free alternative.
Concealer	Glass with metal lid	Can recycle or upcycle container.	Positive.	Find other products with similar packaging.
Mascara	Plastic, varying types	Not recyclable.	I use it all, but negative packaging.	Metal alternative?
Eyebrow pencil	Wood and plastic lid	Shavings from sharpening pencil are compostable; lid is unrecyclable.	Not wasteful if I use it all. Wood element positive.	Find one with a metal lid? Compost shavings.

I will admit that, at first, I was slightly overwhelmed by all the changes I felt I needed to make with my personal care products. There was so much plastic I had never even considered! But over time, I worked through my toiletry products and makeup bag, replacing items and learning along the way. Here are some tips and reminders for this area of your home:

- Instead of using body wash, shampoo and conditioner packed in plastic bottles, switch to loose bars packaged in paper.

- Look for plastic-free and refillable options for all bathroom products.

- Where possible, make your own products, and upcycle glass jars and metal containers to store them.

- Switch to a bamboo toothbrush and compostable dental floss in plastic-free packaging.

- Send any nonrecyclable pieces of packaging (for example, the plastic pumps from bottles of face oil or cleanser) to TerraCycle, a private recycling business that works to transform commonly nonrecyclable materials into new material for products.

- Switch to a plastic-free razor. I love my Leaf razor.

MENSTRUATION PRODUCTS

On average, a person who menstruates has an estimated 450 periods during their lifetime and goes through 20 tampons per cycle. This results in approximately 9,000 tampons used in a person's lifetime. And because tampons and pads are made from and packaged in varying materials, a lot of which involve plastic, they will all go to landfill. Menstrual products can also cost an estimated £4,800 in a lifetime and can be difficult to access for some people. When I first read these statistics, I was shocked! So I looked into reusable alternatives that, if cared for correctly, can be used for life. I found the best advice from asking friends and reading customer reviews to compare the options. If your health allows you to, I encourage you to look into some of these alternatives:

- Period underwear that can be machine-washed

- Machine-washable pads that attach to underwear

- Menstrual cups, which are affordable, widely available and can be used over and over

- Applicator-free tampons used with or without a reusable applicator

TOILET PAPER

Last but definitely not least is toilet paper and tissues. On average, a person uses one hundred toilet rolls in a year. Toilet paper is typically made with harmful toxins such as chlorine bleach, formaldehyde, BPA, BPS and paraffin wax, which are used to make it soft and bright white. Do we really need thick, silky, bright-white toilet paper at the cost of the planet and our health?

- ⊙ If possible, buy toilet paper made from bamboo. Processing bamboo into paper generates 30 per cent fewer greenhouse gases than virgin wood; bamboo is naturally softer; and the plant itself is a renewable resource (it is the fastest-growing plant in the world and can be replenished quickly). I personally love and use PlantPaper.

- ⊙ Use less. Without realizing it, most of us take way more toilet paper than we need each time we go to the bathroom.

- ⊙ Use a handkerchief or other washable, reusable cloth for blowing your nose.

WORKSPACE

Our workspaces all look very different; you might work perched on the couch, at a home desk, at an office, at a studio, at a retail store, out of a rucksack or in a vehicle. The workspace is an environment that these days relies heavily on technology, an industry that produces a lot of harmful emissions and materials. Let's take a look at some of them.

ELECTRONIC DEVICES

First, there are the electronics – laptops, computers, printers, keyboards, mice, drawing pads, tablets and more – all of which can quickly become obsolete and need replacing. One big thing to think about here is how you dispose of them.

⊙ First, can you fix it instead of disposing of it? There are some great resources such as ifixit.com that can teach you how to fix almost any electronic device.

⊙ Is it recyclable?

⊙ Can you donate or sell it?

⊙ When purchasing, can you buy one secondhand or refurbished?

DATA AND STORAGE

A surprising amount of carbon emissions can be expended during your daily tasks and data storage. The information and communications technology industry accounts for about 3.7 per cent of global greenhouse gas emissions annually. This equates to approximately 414 kilograms per person per year. Positive shifts you can make to lower your impact include:

- ⊙ Opt for a conscious cloud provider that is powered by 100 per cent renewable energy. Big tech companies such as Apple, Google and Facebook have taken steps to green their clouds, but I encourage you to do your research.

- ⊙ Unsubscribe from newsletters you are no longer interested in.

- ⊙ Limit 'reply all' or talk in person when possible.

- ⊙ Go through your cloud storage and delete unwanted files and photos.

- ⊙ Use an alternative search engine such as Ecosia, which donates 80 per cent of ad revenues to organizations and community projects working to regenerate forest communities and protect biodiversity hotspots.

PAPER, PENS AND OTHER OFFICE SUPPLIES

Other items that could appear on your desk are stationery, notepads, planners, binders, file boxes, printers and desk lighting. There is so much stationery in the world, a lot of which is made with materials that are nearly impossible to recycle in your local facilities. Here are some small changes you can consider making:

- ⊙ Opt for a classic wood pencil or a refillable pen.

- ⊙ When buying planners, notebooks, binders and filing systems, consider materials that go into each product (Can you find a planner made from recycled paper?) and the product's afterlife should you dispose of it.

- ⊙ If you work in a shared office space or within a company, suggest a private recycling company such as TerraCycle for hard-to-recycle items such as ink cartridges and pens. I was actually first introduced to TerraCycle after seeing it in action at an office.

LAUNDRY AND CLEANING

Cleaning and laundry products keep the house clean but can create a lot of waste. The products we use in and on our home are directly linked to our own health, as we breathe in and even consume the ingredients they are made of. The materials in the packaging will affect our health as they pollute waterways and the earth. A variety of products have been introduced that help to reduce single-use plastic packaging, including reusable dryer balls, dissolvable cleaning tablets you dilute with water and plastic-free dishwasher pods sold in cardboard boxes (I use Dropps). There is also an opportunity here to make your own products. While you might not want to dive into laundry detergent first, you could start with a brightening solution for your white clothing (see page 192). First let's take stock of five items commonly found in the utility room – see opposite.

Here are some things to think about when addressing your laundry and cleaning practices:

- ⊚ Can you streamline your products so it is less overwhelming to find lower-impact alternatives? For instance, a multi-purpose spray can be used for lots of things.

- ⊚ If plastic-free alternatives are not available to you, buy products with a higher concentration so they last longer and can be diluted. Buy larger quantities so the product-to-package ratio is higher.

- ⊚ No more paper towels or disposable cleaning wipes! Instead, use a kitchen towel or fabric napkin that can be washed and last a lifetime.

CLEANING PRODUCT INVENTORY TABLE

ITEM	INGREDIENTS & PACKAGING	IMPACT	POSITIVE OR NEGATIVE	SOLUTION
Laundry detergent	Biologically friendly ingredients. Plastic container.	Ingredients break down in water. Will the plastic be recycled? Harms ocean.	Partly negative – want to change.	Plastic-free alternative, e.g. Dropps pods or powdered soap in a cardboard box.
Dryer sheets	Harmful ingredients. Made of polyester with a softening agent.	Bad for health. Wasteful, goes to landfill.	Negative – want to change.	Do I need them? Find alternative, e.g. wool dryer balls.
Multi-purpose cleaning spray	Toxic ingredients. Plastic packaging.	Breathe in ingredients. Plastic container will not be recycled.	Negative – want to change.	Find plastic-free alternative and biologically friendly ingredients. Or make my own (see page 213).
Paper towels	Paper could have bleach or harmful softening chemicals in it. Plastic packaging.	Wasteful and expensive. Not recyclable, goes to landfill. Plastic wrapping will never decompose.	Negative – want to change.	Reusable, washable cleaning cloths. Cut up old towel.
Washing-up liquid	Biologically friendly. Packaged in plastic.	Plastic might not be recycled. Bad for ocean and land health.	Partly negative – want to change.	Find plastic-free alternative, either refillable or dish soap bar. Or make my own (see page 211).

OUTDOOR SPACES

An outdoor space could be anything from a patio, a balcony, a window ledge, a garden, to a larger piece of land. This is a space that can not only provide joy and fresh air but also an environment in which you can grow edible plants, vegetables and fruit. Here are some ideas for ways to get the most out of your outdoor space, no matter how big it is:

- Consider setting up an outdoor compost unit, whether it's a worm bin, loose compost pile or just a bucket to collect compost that you drop off at a local community garden.

- Growing your own herbs is a fun and easy way to shorten your supply chain and limit your single-use plastic consumption. Check out page 226 for more tips on getting started.

- Any outdoor space can be a wonderful retreat for rest and relaxation. If you're feeling stressed or overwhelmed, take five minutes to go outside and just breathe in the fresh air.

PETS

Pets, while incredibly adorable and unconditionally lovable, can have some wasteful habits, from chewing household items, having big appetites and daily needing to poo on the pavement. When I adopted my dog, Billy Blue, and cats, Frank and Hugo, I did some research to figure out the most environmentally conscious choices I can make as a pet owner:

- Opt for pet toys made from nonplastic materials, or encourage your animals to love simple, natural toys such as a ball of wool for a cat or rope for a dog.

- When it comes to food, seek clean ingredients from sustainable resources and packaged in steel tins (which can be cleaned and recycled) or packaging that can be recycled through private companies such as TerraCycle.

- Another alternative (which I admit I have yet to try) is homemade pet food.

- For other higher-investment items such as pet beds, opt for natural fibres where possible to avoid microplastics washing off in the laundry.

Synthetic fillers harbour odours and will need to be replaced more often. Alternatively, seek secondhand options.

- ⊘ Last but not least, poop! The choice you are faced with here is the same as for your household rubbish bags (plastic, recycled plastic or bio-based plastic). None of these options is great, but rather than using my bare hands, I have chosen to use bio-based poo bags. In the garden, a scooper or pan and brush are a great way to eliminate using poo bags.

- ⊘ Use natural, biodegradable cat litter instead of clumping clay or silica gel. There are lots of options at pet shops, from casava flour to walnut shells.

I hope this chapter has let some fresh air into your home as you consider the possibilities for change. As this book's title suggests, I invite you to take this journey gently and at your own pace. Do not overload yourself – instead, take on one thing at a time. The climate crisis can be overwhelming, so choosing small tasks and completing them one by one will ease that anxiety and return some of the power to you.

go organize

PLANNING TOOLS TO HELP
YOU ACHIEVE YOUR GOALS

My organizational skills are definitely selective. I like to organize areas of my life that I enjoy, like cooking, whereas I am less enthralled to organize my wardrobe and cupboards. But as I've worked to strengthen my personal and community-based action around the climate crisis, I've realized that, for me, being organized plays a key role. When I first began this journey, the more I learned about the climate crisis, the more troubled I became and the more things I wanted to shift in my daily life. But without tools to help me prioritize and organize, I ended up feeling overwhelmed and anxious. Before I knew it, I was trying to change everything at once and striving for perfection, expending a lot of energy without a lot of direction, which induced burnout and a feeling of failure. I had to take a step back and acknowledge that making these changes takes time, thought and organization. We can't do everything, and we certainly can't do everything at once. We must go gently.

I found it best to start small. To some extent, we must organize ourselves before organizing large-scale change. I don't mean to suggest you spend all your time and energy becoming a completely organized person. Instead, take a moment to reflect on the ways you would like to participate in change and the ways you can realistically achieve this. I think if we can see change unfold in our own life, we can more genuinely believe in the ability for systemic change. For me, the most useful tools have been a habit tracker, weekly planner and checklists to keep me on track with my goals when I'm travelling, socializing and working. They helped me organize my thoughts and goals on both a personal and community level, track and pace my growth, and stay the course when I'm out and about. Have a go with these ideas a few times to get the basics, then adapt them so they work for you.

HABIT TRACKER

To avoid feeling overwhelmed, I started making lists of all the things that interested me under the large umbrella of the climate movement – groups that inspired me, topics I wanted to learn more about, issues affecting my local community and changes to my personal life, such as what I consumed and who I voted for.

By doing this, I began to see that there were two streams of action. One was more personal and intimate – improving how I interact with the planet, such as buying less and composting my food scraps. The second was taking action in my wider community – joining groups working to implement policy at a local and national level or providing access and resources to underserved communities. I knew that I wanted to be part of both these streams of change, so I tried to see how I could balance and stay on track with the two simultaneously.

I love to write lists, but actually staying on top of everything is an entirely different discipline. I'll admit I have even written items on a to-do list that I have already done just so I can tick them off. I'm not sure that is actually productive, but I find a lot of joy and satisfaction in completing tasks. I realized I needed to marry my love for list-making with my desire to commit to taking action, all the while doing it at a pace that was gentle and patient with myself. So I started using a habit tracker, which is a tool to help develop new habits you are working towards on either a daily, weekly or monthly time scale. This can help you track your progress and visualize the change you are making.

The first few times I drew up a monthly habit tracker, these were some of the items I wrote down:

- ⊚ Take food scraps to local community garden compost drop-off.

- ⊚ Say no to plastic coffee cups and take a reusable.

- ⊚ Don't buy fast fashion.

- ⊚ Watch a documentary on the climate.

- ⊚ Sign petitions.

- ⊚ Research and educate myself on where my rubbish goes.

- ⊚ Share something I learned with a friend.

This is what one of my habit trackers looks like:

This may look like quite a formal approach to taking action, and I will admit I create habit trackers less and less now that the habits are engrained. What I found most interesting about this exercise was that I began to identify which habits were the most challenging to implement. When I questioned why, sometimes it was because I was being human and could forget things or be lazy, but more significantly these habits were challenging because of certain systems preventing me from making the choices I wanted to.

For instance, trying to buy less single-use plastic was extremely difficult at first as so many products I was used to buying are packaged in plastic. I found this pretty dispiriting, but my frustration around this and other issues then empowered me to do some research and try to figure out ways to advocate for larger, more systemic change.

My advice with a habit tracker is to take things slow. Don't fill it with new habits that are impossible to implement all at once. Start with just three things and go from there. This is the formula I use, but I invite you to change it and make it your own to best track and celebrate your change and growth.

WEEKLY PLANNING

I'm often asked, 'But how do you find time to cook for yourself?' or 'How do you always remember to bring your reusable utensils to lunch?' It all comes down to planning. I personally love to organize my time and keep a planner and, on reflection, that's because I have had to for so long. From a young age, I was balancing schoolwork while shooting *Harry Potter*. I would go to each one of my teachers at school and get the entire semester's lesson plan and supporting materials and stay on track while filming. Not to mention that I became a teenager during this time and discovered the importance of organizing a social life in between! These days, I'm still juggling work and social engagements, plus advocating for change within my community and while doing life's inescapable tasks, such as the laundry. A weekly plan is now more essential than ever.

I begin by identifying what kind of week I have ahead so that I can best prepare. Thank goodness for the notes and alarm apps on my phone. First, I fill in all of my commitments for the week – work meetings, film shoots, social engagements, culture, exercise and climate rallies, for example. I then think about food, identifying which meals will be at home, on the go or eaten outside of the home. Next, I look for pockets of time to devote to individual and community environmental work, getting outside to clear my head, and household jobs such as making more cleaning spray or reorganizing my wardrobe. I find that if I put things like this in the calendar at a set time, I am much more likely to follow through.

ANNIE LEONARD
Executive director of Greenpeace USA

How can individual action help us understand the climate movement as a whole?

Individual action has a lot of power. It helps us bring into alignment our personal values and our actions; then we feel more integrated. We feel more whole, we feel better, the more aligned we are. I also like to think of the individual actions as a metal detector to find the flaws in the system. Anywhere that it's hard for us as individuals to do what's right, that's a systems flaw, and that helps us find it so we can fix it. We should redesign our consumer economy so that doing the right thing is the easy thing to do. Instead, we should make the people who want to pollute, go out of their way to pollute. The system isn't like that now, and so when it comes to packaging or waste, anything that you can't avoid or you can't recycle, boom! – that's the system's flaw. Individuals can only do so much within a flawed system, so we need to do both: make good choices and work together to change those systemic flaws so it is easier to make good choices.

What is one piece of actionable advice that you would give?

I think that the most important thing to do is find some friends who also want to make a change so that you have support, because we're stronger when we're together. Connect, find a group, find some friends, don't try to do it alone. That's my most important thing. Also, pick the thing that you're interested in – there are so many ways to engage that it's like a huge, diverse buffet of opportunities; there is something for everybody. If you start working on turning vacant areas of space into community gardens and that's not for you, totally cool. There's a billion other things. Keep trying things until you find one that feels right. We don't have to be martyrs; we can make a difference and have fun doing it!

What instills hope in you?

I'm hopeful because change is happening. If you look at public polls about how many people are concerned about plastic waste, how many people are concerned about chemicals in our products, how many people want stronger government action on climate, we have most people with us. That's what we need to drive change, most people. What we're missing, which is why we need your book, is the guidance to go from being concerned to being active.

And by planning, I can attempt to balance things out so there is space for both joy and getting my hands dirty, and putting the work in from week to week.

Some types of things that can end up in your weekly planner might be:

- Times you'll be eating out

- When you need to take a meal on the go to work or an occasion

- What day(s) you will do a food shop

- Time you could allocate to batch cook or prepare meals for the week

- A local climate event or community group meeting

- Time outside or moving your body

- Time for rest or reading, listening to a podcast or watching a film

- Signing petitions or writing emails to your representatives or corporations and brands

- Laundry and cleaning day

- Organizing your cupboards and storage

- Small DIY projects

I like to start my week planner on the day that I will be doing a food shop, as that's usually when I have time to store everything away, stay organized and possibly cook some larger meals to sustain me through the week ahead. This may differ for everyone; it may be more fitting to start your week on the first workday of your week or the day you will find some time to rest. I tend to pack more into the first half of my day because that's when I feel most energetic, but this will also differ for everyone. Think about the times of day when you are most productive, and use that to schedule your tasks. On the next page is an example of an average week for me. I have two layers: the green is the event or the task, and the blue is a note to myself on what to remember or do to best keep in line with my values and the change I am trying to be.

I invite you to use these planners as a guide for creating your own, focusing on the issues you care about and the habits you'd like to change in your life. Remember to be gentle with yourself and start slow, adding manageable tasks that you feel you can realistically accomplish each day.

TIME	SUNDAY	MONDAY	TUESDAY	WEDNESDAY	THURSDAY	FRIDAY	SATURDAY
MORNING	Breakfast picnic with Lucy on a hike.	Breakfast on the go.	Breakfast at home.	Breakfast out with a friend.	Breakfast at home.	Breakfast on the go.	Yoga class.
	Text Lucy and ask what we should both take. Could she pick me up to not take two cars?	Take a homemade breakfast bar and coffee.		Find a café to dine in.	Email my community group in response to this month's campaign.	Prepare overnight oats the night before.	Take mat, reusable water bottle and coffee cup.
MIDMORNING	Food shop at farmers' market and supermarket.	At a shoot.	Work from home.	Work from home.	Gym.	Scouting location for short film shoot.	Coffee with friend.
	Take reusable shopping bags and write shopping list.	Email producer and ask about the catering. Reusable utensils and cup?	Between work, email friends of the Greenpeace Oceans Treaty petition.		Take banana, reusable water bottle. Tote bag for sweaty clothes.	Pack lunch, utensils, water bottle and sunscreen.	
NOON	Laundry and tidy house.	At a shoot.	Lunch meeting.	Work from home.	Production meeting for short film.	Scouting continued.	Picnic with friends at the park.
	Check cleaning supplies and maybe make more.		Check if it's a dine-in restaurant.		Bring up waste management to the producer.		Go via bakery and pick up bread; take homemade dips.
AFTERNOON	Week food prep.	At a shoot.	Coffee meeting.	Check up on garden.	Food shopping for dinner party.	Meet with crew at coffee shop.	Home to work on Monday's presentation.
	Make roast veggies?		Take reusable cup.	Time to plant seeds yet?	Text friends – who is cooking what?	Take reusable cup or find a dine-in café.	
EVENING	Dinner at home. Prep for shoot.	Spanish class.	Dance class.	Attend a talk with friend, then dinner.	Book club.	Dinner out and friend's house party.	Dinner out followed by cinema.
	Watch that documentary on seeds!	Take laptop, workbook, dinner and drink.	Dance clothes and water and snack.	Find somewhere near talk we can walk to.	Take wine and homemade beetroot dip.	BYO cup and wine.	Find a restaurant that cooks using local produce. Take chocolate for film.

TRAVEL, WORK AND SOCIALIZING: HOW TO STAY ON TRACK WHEN YOU'RE ON THE GO

As you begin or continue to implement positive change in your life, experiences present themselves, such as travelling or attending a work event or social occasion, and suddenly your grasp on these new habits can crumble. First, there is absolutely nothing wrong with that – we need to live and go with the flow. But there are a few ways to equip ourselves so that we can continue to live our values in new situations.

TRAVEL

I often travel for work, and over time I have built myself a travel kit that helps me to be less wasteful and more self-sufficient. It's been a long time since I have used a miniature shower gel bottle in a hotel, and I love that I can keep my products consistent when I travel. I have even been known to bring washing-up liquid and a laundry detergent pod or two when I go away on longer trips or stay in a rental property. Your travel kit may look different – it will reflect the things you love and the issues you care about – but I hope mine helps get you started. The kit falls into two categories: food and bathroom.

My Travel Tool Kit

Food and Drink

- ⊙ Reusable water bottle. This keeps me from drinking out of single-use plastic on the plane or when I'm out and about. I can fill it up at public drinking fountains, hotels or cafés.

- ⊙ Reusable coffee cup. This gets a lot of use! I bring my KeepCup to meetings at coffee shops and fill it up on set.

- ⊙ Bamboo utensils. These keep me from using disposable plastic utensils for meals on the go. I love my To-Go Ware set, and so far all airport security has allowed me to travel on board with it. For daily use or on road and train trips, a regular set from your kitchen works perfectly, too.

- ⊙ Grocery shopping bag. So I don't need to use the plastic ones when I'm picking up fresh produce for snacks or cooking. I also find these great to organize and store items in while I am packing.

- ⊙ Reusable silicone storage bag. Perfect for packing up leftovers or snacks on the go.

- ⊙ Tote bag. In case I find anything when out shopping or need to pack up a change of clothes for a workout.

- ⊙ Headphones for the plane. To save having to use the airline's often disposable ones.

Bathroom

- ⊙ Soap box with soap. I love my little metal box I bought at a charity shop. There are also ones you can buy online.

- ⊙ Cotton washcloth. This saves me from using disposable makeup wipes. I can easily wash and dry it as I travel.

- ⊙ I decant my own products, including shampoo, conditioner, face cleanser, moisturizer and toothpaste, into small bottles. There are also great shampoo and conditioner bars on the market.

If you travel a lot, try to find hotels that have a more sustainable approach to products, services and catering options. When looking into rental properties, I like to ask the host what they offer in terms of bathroom and cleaning products and even their kitchen utensils and appliances before booking. This information can help you identify what to bring yourself. I love coffee, so if they only offer a coffee machine with disposable pods, I like to bring either my mini camping titanium French press or my reusable fabric coffee filter. Now, I am not suggesting you pack your kitchen sink (even though I would secretly love to sometimes), but it's nice to know what to expect before you arrive so you have the chance to bring a few of your favourite items and be as resourceful as possible.

SOCIALIZING

When I started to adopt new habits to consume less single-use materials, part of me dreaded socializing. I was nervous about navigating single-use plastics at picnics or looking strange if I pulled out my reusable mug at a coffee shop. But I quickly realized that, first, no one cares; second, I might inspire friends to also change their habits; and third, I really only needed a few things in my bag to cover any given situation. I now carry most of these items whenever I go out so I'm ready for anything and can be prepared even if plans change.

My Socializing Tool Kit

- Bamboo cutlery (or a set of utensils from home). You never know when ice cream might be on the cards after dinner!

- Reusable coffee cup.

- Reusable cup. I bring this for picnics or when I'm going to a party where I think they might be serving drinks in disposable and nonrecyclable cups. I have a very loved one I got aboard the Greenpeace ship *Arctic Sunrise*, which is a Klean Kanteen cup. This is also great for camping.

- Cotton cloth napkin or handkerchief. Better than any disposable paper napkin!

WORK

As a filmmaker and environmental advocate, I don't have a typical 9-to-5 workday or office setup. I'm usually either working from home, in meetings or on set. In some ways, my freelance schedule makes it easier to keep in line with my values (I can often make lunch at home, for example), but it also presents its own challenges (there can be a lot of waste on a film set). To help me stay on track, I've devised a checklist of things to keep in mind while I'm working. Depending on the type of work you do and which environment you work in, your list will look different, but I invite you to think about which small changes you could make and write them down. Here's my list:

- ⊙ Always take my reusable water bottle and coffee mug on set.
- ⊙ Take reusable mug to coffee meetings.
- ⊙ Pack up breakfast and lunch in reusable containers and take bamboo cutlery when I'm going to be out all day.
- ⊙ Look into more sustainable catering options on set.
- ⊙ Take homemade snacks for long days.

Incorporating these organizational tools into my life has helped me significantly reduce my stress, increase my productivity and achieve the goals I have set for myself. Of course, there are still times when I forget my mini shampoo bottles on a trip or use a disposable coffee cup at a shoot. In these moments, I try to remember that we can't control everything, and the minute we beat ourselves up, we distract ourselves from feeling confident about change. So as you work towards becoming the most organized, climate-conscious version of yourself, I urge you to record and celebrate your achievements and not beat yourself up about any small hiccups. We must go gently.

go shop

HOW OUR CHOICES CREATE CHANGE

This has been a challenging chapter to write because shopping with the planet and humanity in mind is incredibly complex. We can pick up an item and question if buying it is 'good' or 'bad' for the planet, but for all those existential moments I have had at the supermarket or hovering over a buy button while trying to be a better human, the answer is never that simple.

Over the years, I have found that as soon as I stopped categorizing my choices as right and wrong and instead began looking into all of the nuanced factors involved in getting the product into my hand, the calmer and more empowered I felt. My decisions now feel more informed and intentional. It can be tedious to ask brands questions, look up information online and learn more about materials and resources, but it is our responsibility – to the planet that gave us these resources and the people who shaped them – to be curious and conscientious.

BIG-PICTURE BUYING GUIDE

I can't say if any particular purchasable item is right or wrong, but what I can offer is the guide I developed to help navigate my own choices, which contains information on materials, marketing language and certifications that I've learned over the years. My Big-Picture Buying Guide helps me think about the whole supply chain involved for any given product, including:

① Raw materials and ingredients

② Growing and extracting

③ Social justice and impact on livelihoods

④ Processing and manufacturing

⑤ Packaging

⑥ Shipping and transportation

⑦ Product use

⑧ Afterlife of the product

To put this into practice, let's take a simple loaf of bread through all eight stages of assessment. I am a visual person, so I like to keep this illustration saved in my phone to return to for those moments when I slip back into thinking that there is only the right-or-wrong, good-or-bad way to look at an item.

BUYING GUIDE (pages 94–95) ILLUSTRATION KEY:

① Raw materials and ingredients include salt, wheat, trees, fossil fuels and water.

② **Growing and extracting:** Salt is harvested from evaporated ocean water, trees are grown and logged, wheat is grown and harvested, fossil fuels are extracted and water is taken from the local water supply.

③ **Social justice and impact on livelihoods:** At every stage of production, people are involved, and their health, safety and access to a living wage is impacted.

④ **Processing and manufacturing:** First the harvested plants and minerals are processed into the ingredients to make the bread. The salt is ground, and the wheat kernels are milled into flour. The packaging materials are made: tree pulp is processed into paper for the flour bag, and petrochemicals are processed into low-density polyethylene (LDPE) for the plastic packaging for the salt and bread. The bread itself is made through the process of fermentation and baking.

⑤ **Packaging:** The bread is packaged in an LDPE plastic sleeve with a plastic tie and then packaged in cardboard boxes to be transported to the supermarket.

// continued

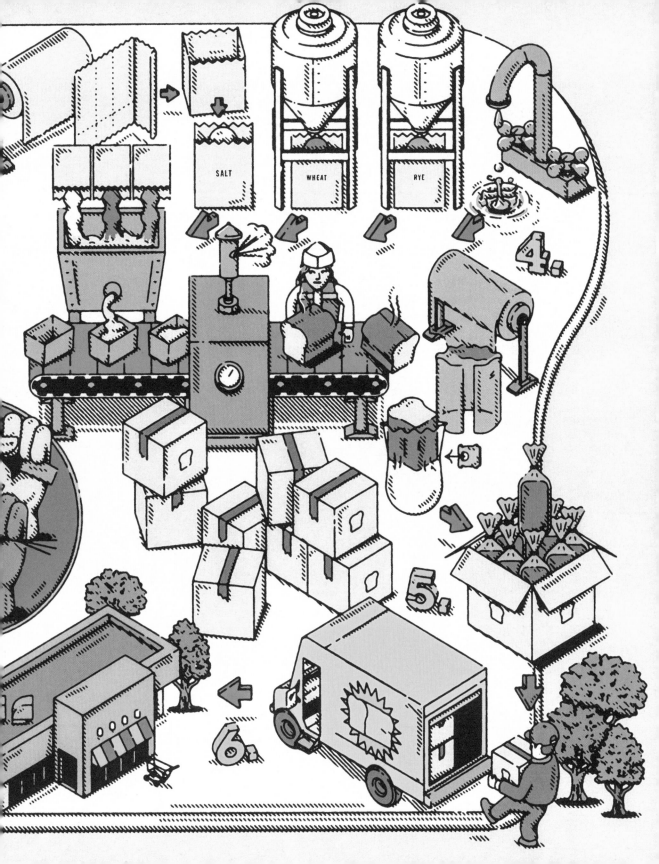

SALT

WHEAT

RYE

4.

5.

6.

6. **Shipping and transportation:** This happens between almost every stage of the bread's production, from the ingredients being transported to where the bread is manufactured, where it is packaged, where it is sold, how the buyer takes it home from the store and finally, from curbside to landfill.

7. **Product use:** The bread could potentially be toasted, which uses electricity.

8. **Afterlife:** If you don't eat all the bread, some could go to landfill (or, ideally, composted). The plastic packaging is likely to be unrecyclable, so this too will go to landfill.

By taking one product through this cycle of review, you can now apply this set of questions to any item you're considering buying. You will find during your inquiry into all these entangled layers that it is sometimes almost impossible to make your preferred choice, and this often exposes where a system is purposefully working against us. For instance, if an essential product is only available in single-use plastic packaging. This is where we simultaneously have to compromise and also dig further into why that system is broken and needs reimagining to make it easier for us to make better choices. Our choices are mainly based on cost and availability, so compromises have to be made, but we can still challenge manufacturers and shopkeepers to offer products with planet health in mind rather than only commercial expediency.

CLEAN-UP ON AISLE 5

Now that we have visualized the layers involved in the production of a product, let's look at the way that the product is marketed to us. This could be the language the brand uses in marketing campaigns, the ethos advertised on its website or what's printed on the packaging and labels. There is a broad spectrum of extremely powerful language here, and I find myself easily misled.

As I work to make shopping choices that better reflect my beliefs, I have realized that I have to assume more responsibility instead of blindly accepting that what a brand prints on a label is true. I think there has been a general misuse of language in the past few decades and a certain artistic licence that strips the true meaning from the word. Your definition of 'sustainable', for example, might not be someone else's, and companies are constantly taking advantage of these grey areas to try to sell us products. As consumers, we must be active, curious and diligent in our choices. Here is a breakdown of commonly used terms, labels and certifications to help you navigate the aisles.

GREENWASHING – DON'T LET IT FOOL YOU

We are constantly bombarded by greenwashing. This is when companies and even the government use language that suggests they are doing more to protect the environment than they really are. Before we even think about what is in a product or what it does, we are more often than not greeted by greenwashing language. Let's unpack some of these buzzwords:

'Sustainable'

An incredibly misleading and overused term. Can we really call any type of shopping sustainable? The best definition of 'sustainable' I've heard is, when looking at the development of products, a product that meets the needs of the present without compromising the ability of future generations. The reality is, most products, processes and consumer goods are compromising our future generations tremendously.

If a company claims they are 'sustainable', research and question what exactly is sustainable about them. Is it the materials the product is made from, the packaging, the processes used to make the product or the afterlife of the product? If a brand calls itself sustainable, it should be addressing the impact of all stages of the production and afterlife of a product. We cannot let 'sustainable' become a marketing buzzword and completely lose its meaning. Otherwise, this disrespects those doing true sustainability work.

'Green'

I get lost with this one. Technically, saying something is green is telling me absolutely nothing. It's a vague word that implies the product is somehow environmentally friendly, but the truth is, it doesn't mean anything. Notice, too, how the colour green is used on packaging to suggest that a product is good for the earth or is from the earth. For many people, the colour green has a strong association with nature and conjures up images of lush trees and open spaces. Brands cleverly use the colour green in label and logo design and in adverts to quickly transport us to our stereotypical perceptions and create the impression they are a safe choice.

'Natural'

Everything on Earth, in theory, is natural, but after any given ingredient or material is extracted, a plethora of chemical processes and modifications can take place that take the virgin ingredient very far away from its 'natural' form.

Research by Consumers Union found that 86 per cent of consumers understand the term 'natural' to mean a product that doesn't have any artificial ingredients. The best thing you can do is look closely at the ingredients list and see how many items you truly recognize as whole and clean ingredients. Every brand will have a different definition for natural, so don't assume it shares yours.

'Cruelty-free'

What we think a brand is telling us by saying it is 'cruelty-free' is that none of its products, materials or ingredients were tested on animals. However, without certifications and detailed language that confirms it can guarantee no animal testing was done, including by any other companies it works with to create a product, it's hard to know whether a product is truly cruelty-free. If the brand says, 'We don't test on animals', that could still mean it buys ingredients from other companies that do or that it pays others to do the testing for them. Sometimes brands just print a bunny emoji–type design on their packaging and say 'cruelty-free', because the actual Leaping Bunny certification associates the image of a bunny with the idea of cruelty-free. If it doesn't look like the official certification, don't trust it. Because the term 'cruelty-free' is ambiguous and there is no legal definition for what this means, a better question to ask a brand is, 'Are your products tested on animals?'

○ ○ ○

Overall, the issue with language used in blurbs, labels and marketing materials is that it gives brands far too much freedom. False claims can be challenged in court, but until these brands back up their 'sustainable' claims with hard science and facts, they can tell you just about anything. Try as much as you can to ask questions, not assume, seek transparency and hold brands accountable if you think they are misleading you.

LABELS AND CERTIFICATIONS: WHAT DO THEY ACTUALLY MEAN?

A lot can be printed on a package, a label, a billboard, a website, but what does it all mean? I sometimes find myself staring at a package, decoding the official and unofficial labels and still not knowing what they really affirm. We assume that labels have been designed to help us identify what is what, but some labels can be designed to be intentionally misleading. Take, for instance, plastics codes. Plastic types each have a number, and these numbers are printed within the chasing arrows symbol that we commonly associate with recycling. This makes us believe that the package is recyclable or made of recycled materials. Actually, all it's telling us is that it's plastic and it's a particular type of plastic. Let's break down some of the materials labels we encounter on a daily basis.

MATERIAL TYPES: CAN I RECYCLE THIS?

Plastics

There are seven types of plastic, numbered 1 to 7. Plastics are made from fossil fuels, tying them to one of the most damaging industries on our planet. Companies make money by producing and selling plastics, so their successful business model is built on the idea that we will dispose of plastics and need more. They do not make money if we are able to create a fully closed-loop recycling system or, better yet, not use plastics at all.

Claiming that plastics are recyclable is somewhat untrue. I often get asked questions like, 'Is this plastic recyclable?' or 'Does any of my plastic recycling ever get recycled?' And my answer is, not really. To understand why, let's first remind ourselves of the definition of recycling: taking something, breaking it down and turning it back into the same thing. We also have downcycling, which is taking something, breaking it down and turning it into something else. Downcycling sounds more like the reality of what happens when we recycle plastic. For example, a water bottle made of PET could be recycled into nylon fabric for clothing. It won't be made into another water bottle because beverage companies can make cheaper new bottles with virgin plastic. Once plastic becomes clothing, it can't be recycled. The more we as consumers demand products made from 100 per cent recycled materials and not virgin resources, the more money corporations will spend on developing products from 100 per cent recycled materials and the more valuable recycled and recovered materials will become.

Here is a breakdown of plastic types:

CODE OR LABEL	WHAT IT IDENTIFIES	WHAT IT'S USED FOR	CAN I RECYCLE IT?	TAKEAWAYS/SOLUTIONS
01 PET	Polyethylene terephthalate plastic	Water bottles, condiment bottles or jars	Generally accepted by curbside recycling programmes.	Use a reusable water bottle. Never heat PET, as it causes toxic chemicals to leach. Rinse out before recycling. Leave plastic lids on water bottles when recycling (if separated, they often get lost and end up going to landfill).
02 PE-HD	High-density polyethylene plastic	Plastic milk cartons, shampoo bottles	When recycled, turned into items including new shampoo bottles, but most require additional virgin plastic to make.	Make your own plant-based milk or buy milk in glass bottles. Switch to shampoo bars. Rinse out before recycling.
03 PVC	Polyvinyl chloride plastic	PVC pipes and siding, medical equipment, children's toys	This is rarely recycled and highly toxic to both humans and the environment.	Do not put this in your curbside recycling, as it will not be recycled. Avoid if at all possible.
04 PE-LD	Low-density polyethylene plastic	Plastic shopping bags, squeezable bottles, bread bags, frozen food packaging	If recycled, this can be made into bin liners and plastic lumber.	Rarely recycled. LDPE items such as plastic shopping bags have to be picked out by hand at the materials recovery facility (MRF). Take reusable bags to the supermarket.
05 PP	Polypropylene plastic	Yoghurt containers, syrup bottles, prescription bottles, caps and straws	Generally accepted by curbside recycling programmes.	Use metal or glass straws instead of plastic. Buy yoghurt in bulk, in glass containers or make it yourself. Avoid stuffing items inside each other and then recycling them.
06 PS	Polystyrene plastic	Food takeaway containers, disposable plates and cups	This is rarely recycled.	Use a reusable coffee cup. Take your own storage container for leftovers at restaurants.

CODE OR LABEL	WHAT IT IDENTIFIES	WHAT IT'S USED FOR	CAN I RECYCLE IT?	TAKEAWAYS/SOLUTIONS
△ 07 O	Miscellaneous and polycarbonate plastics	Found in plastics that contain bisphenols (e.g. BPA). Used to make baby bottles, sports drink containers and cosmetic packaging.	Because these are often an unknown mix of plastics, they are generally not able to be recycled and are not recommended for reuse.	Avoid as much as possible.
△	Recycled content	Found predominately on plastic and paper products to signal the amount of the product made from recycled materials.	Refer to the plastic number to know. If a paper product, yes.	Buy 100 per cent where possible.

Bio-based Plastics

These materials are booming onto the marketplace, and it's easy to assume, as I did, that this is a positive step towards solving our pollution epidemic. Sadly, it isn't that simple. Most bio-based plastics perform in a very similar way to fossil fuel-based plastics and cannot ever truly break down and return to the earth. For a bioplastic to bear that label, at least 51 per cent must come from biorenewable resources (it doesn't need to be 100 per cent). The biorenewable resources may include cornflour (genetically modified or not), glucose or byproducts such as seaweed, wood or food waste. The process of turning these materials into something such as an iced coffee drink cup can be highly energy intensive. And what do we do with this cup once we've enjoyed our coffee? While some bio-based plastics can be recycled (it's rare, so look if your local council accepts #7 plastics), very few will ever be 'composted in industrial facilities' as their labels suggest. These industrial facilities are not only hard to access but rarely accept items such as iced coffee cups. So instead, the only place to put these bio-based plastics is in with the general rubbish. They will then head to landfill where they will still take hundreds of years to break down, if ever, as, like plastic, their properties make them incredibly resistant to biodegradation.

Where possible, the better solution over bio-based plastics is refillable containers – that could be your reusable coffee cup, utensils, food containers and reusable produce bags. For something such as a coffee or a snack, could you take a moment to sit down and consume it using totally washable china or glassware?

Watch out for claims such as '100 per cent compostable' or '100 per cent biodegradable', as, similar to many greenwashing terms, these statements are not regulated. Your idea of compostable and biodegradable and the dictionary definition may be different from that of the company selling it to you. I always ask myself, Does this feel and look like plastic? Would my plants easily eat this? If the answer to the second question is no, then it likely has a complex chemical structure and won't be beneficial to the earth. If it's soft and breaks apart easily in my hands, perhaps this can be composted in my home composting system or garden waste bin. Products like these should have a 'home composting' certification or labelling.

Paper, Cardboard and Wood

Labelling on these three materials has crossovers as they all derive from trees. Paper and cardboard can be made from either virgin tree pulp or recycled paper. Buying products that have been made with recycled paper cuts down on emissions and decreases the number of trees being cut down. According to research in the UK, we use approximately 9.9 million tons of paper annually and recycle around 80 per cent of that.

Another layer to paper packaging is that it can sometimes have a plastic lining. A lot of food packaging is lined with plastic to help prevent oils in the food from soaking into the paper. However, once a material is mixed and not separable, it becomes challenging to recycle. There are some programmes that are beginning to address this but not yet at a national level. For wood products, buying items made from reclaimed wood is like buying something made from recycled materials – the wood has been diverted from landfill and given a second life, cutting down on the demand for newly sourced lumber.

Here is a breakdown of labels you may find on packaging:

CODE OR LABEL	WHAT IT IDENTIFIES	TAKEAWAY/SOLUTIONS
Widely Recycled	It is commonly recycled. 75 per cent of the collected materials are recycled by local authorities.	If the packaging has a variety of materials, you will often see the label identify which parts can and cannot be recycled. Or, whether you should separate the materials, such as a metal lid from a glass jar.
RINSE Widely Recycled	Needs rinsing before you recycle it.	Make sure to rinse it!
Not Yet Recycled	This label informs you that this packaging cannot be recycled.	Either upcycle it, find another use for it or put it in your general waste bin.

Glass

Next, we have glass, which can be recycled forever. The recycling process for glass is actually less energy intensive than for metals, and making virgin glass requires a lot of energy, so let's recycle that glass and keep it in the system!

One other important thing to be conscious of with glass is its weight. If the item has travelled a long way to get to you, that will result in a higher output of greenhouse gas emissions. One item I love that is packaged in glass is wine, so I try to buy wine produced and bottled as local to me as possible.

CODE OR LABEL	WHAT IT IDENTIFIES	WHAT IT'S USED FOR	CAN I RECYCLE IT?	TAKEAWAYS/SOLUTIONS
	Recyclable glass material	Condiment jars, wine bottles, olive oil bottles, etc.	Glass can be recycled forever, but some places have stopped collecting it because it is so heavy and therefore expensive to transport.	Try to buy products packaged in glass as locally as possible to reduce transportation greenhouse gas emissions. Upcycle glass jars instead of recycling. Before recycling, check with your local council to make sure they accept glass. Separate lids from glass containers.

Aluminium and Steel

The more research I have done into recycling and materials, the more I have fallen in love with metals. My parents are jewellers, so I witnessed early on how precious metals can be made into something beautiful time and time again. Imagine if we treated the more everyday metals such as aluminium and steel with the same high value as gold. I truly believe that our future will be entirely about recovering the resources we have discarded as the only way to build new things. When I toured a materials recovery facility, the 16-cubic-foot bricks of condensed aluminium cans were reaching as high as £4,000 value to recyclers, the most valuable material at the facility.

Waste = Energy

'Waste' is not the end of a cycle. By acknowledging this fact, we dismantle the concept that waste isn't our problem and are forced to take ownership over it. Now that these momentarily discarded resources are ours, we can find and design new homes for them. This process can create entire industries and jobs. By viewing everything as a resource and energy, we can see it as a potential new source of life. We truly close the loop.

CODE OR LABEL	WHAT IT IDENTIFIES	WHAT IT'S USED FOR	CAN I RECYCLE IT?	TAKEAWAYS/SOLUTIONS
(alu)	Recyclable aluminium material	Drinks cans, aluminium foil, foil trays, etc.	When rinsed properly, pure aluminium can be infinitely recycled. Fits the true definition of recycling, as an aluminium can will become another aluminium can. Doesn't require any new materials (which makes it a closed loop system).	Aluminium is a precious material that needs to be recycled. Clean and empty your aluminium items before recycling. If a can has a plastic sleeve label, remove it before recycling.
(fe)	Recyclable steel material	Tinned beans, soup, sweetened condensed milk, etc.	Items such as a steel food tin can be recycled forever, with no virgin steel added.	Remove any paper labels and rinse steel tins before recycling. Look for BPA-free certification (BPA is linked to reproductive issues, heart disease and diabetes, among other health conditions).

Standing next to a 16-cubic-foot brick of aluminium cans at EDCO Materials Recovery Facility in San Diego.

DR MARTA PAZOS

Chemist, polymer scientist and product and packaging developer

For many consumers, it's incredibly challenging to figure out what can and can't be recycled. In your opinion, how much responsibility lands on the consumer versus the business when it comes to the afterlife of materials?

I think that as humans, we need to stop thinking about corporations and consumers as separated entities. Corporations are made of consumers; consumers work in corporations. We need to stop putting responsibility elsewhere and start taking it ourselves. I have not yet met a straw that goes to the beach by itself. As consumers, we need to be thinking, 'What can I buy so I can do less damage when I have to dispose of it?' or 'What can I *not* buy?' And for corporations, it's not necessarily to pick up the slack or pick up the rubbish, even. It is to facilitate a way for the consumer, without even thinking, to do the least amount of damage.

What is one piece of actionable advice you would give?

Be responsible for what you're doing. Responsibility. Don't think that somebody else will do it for you, because they will not. Be thoughtful with what you buy, be thoughtful with what you throw away. Think twice before buying another pair of shoes. Do you really need that pair of shoes? I'm telling you, I've been shopping in my closet this past year and it's been a lot of fun.

What instills hope in you?

Probably the thing that gives me the most hope is that I feel that our planet is so fantastic that it has continued to find a way to regenerate, and by giving it a little breather, a lot can be done. I do believe that the more people are actually exposed to the truth, the more we'll realize that something needs to change and that change can start with you. It goes back to responsibility; a little bit of effort goes a long way with the earth's ability to regenerate.

CERTIFICATIONS: WHAT DO THEY REALLY CERTIFY?

We've covered greenwashing and the complex world of materials labelling, now let's talk about certifications. The supermarket shelves can feel like an overwhelming sea of certifications these days (Organic! Fair Trade! Leaping Bunny!), and while these certifications can be extremely helpful information, they are not always what they seem. Because third-party certifications are not regulated by the government – they set and regulate their own rules – it is all about how much we trust them.

I do think that certifications can be a great way to navigate retail choices because it helps you to understand a brand's morals and practices, but there is also an inherent grey area within them. Certification bodies make a profit from people using their standards. It's hard to be 100 per cent sure that they are checking that a brand is following the rules and regulations. I would encourage you as always to be curious with your purchases. If there is a brand you love to buy from often, look a little deeper into what it stands for as a company and see if it's in line with what you think is right. Send an email to get clarification on what its standards or use of certifications guarantee.

There are so many certifications out there. I have rounded them up into categories and included the ones that I have personally come across. This list is just the beginning. I encourage you to look into each of these and see for yourself how these certification organizations present their standards.

MATERIALS

The certifications found on materials can tell us all kinds of things, including the afterlife possibilities of the material (such as if and how it can be composted), whether the material contains harmful ingredients such as bisphenol A (BPA), or in what conditions the material was grown and harvested (such as the FSC certification for wood). While many of these are worth seeking out, it's important to remember that not all certified materials are certifiably good for the earth. For instance, I used to quickly read the large print on Biodegradable Products Institute (BPI) products and assume that they were 100 per cent compostable. It was only when I read the slightly smaller print that I realized these materials can only be broken down in special industrial composting facilities. Unfortunately, these facilities are rare, and many won't accept most types of bio-based plastics, which means these materials usually end up in landfill. What I now look for is packaging that says 'home compostable', as that means the material can decompose in my home compositing system. So, as always, do your research.

FOOD

Food labels and certifications are some of the most important to me. Not only do they shed light on what I might be putting into my body – after all, if the food I eat was grown with carcinogenic chemicals, there's a good chance I'm also ingesting them – but they can also give insight into things such as the living conditions of animals and working conditions for farm labourers. I personally look for 'Soil Association Certification' and 'OF&G Organic Certification' when buying produce and pantry items and 'RSPCA Assured' and 'British Lion Quality' for eggs. One type of certification I've learned not to trust is for seafood, as the fishing industry can be very corrupt and difficult to navigate, making these certifications some of the least monitored ones out there. It is challenging to really know if the fish was 'sustainably caught'. My advice is to buy directly from a small-scale fishmonger and eat seafood less frequently. A helpful website to use belongs to the Marine Conservation Society, which offers a wealth of information and ratings based on stock status, management and capture techniques.

HOUSEHOLD CLEANING PRODUCTS

Cleaning products such as dish soap and laundry detergent are often full of chemicals that can be harmful for our health and the planet. Your home is an incredibly personal space, and the products you use on your surfaces and on your clothes have an intimate relationship with your body. A lot of these products also enter our water systems through the drains. These drains lead to the ocean, and this pollution will eventually make it back to us. Look for certifications such as EU Ecolabel, and, where possible, buy products with ingredients that are the least harmful.

BEAUTY AND HYGIENE PRODUCTS

Similar to household cleaning products, we have an intimate relationship with our beauty and hygiene products as the ingredients interact with and are absorbed by our skin. Our skin is the largest organ in the body, so what we put on it is as important as what we eat. Look closely at the ingredients that are in your products, and favour more natural and fewer chemical ones. I tell myself that if I can't pronounce or recognize an ingredient, it probably isn't that healthy for me. I often notice the 'Ecocert' and 'Soil Association' labels on beauty products I buy in the UK, both of which are under the 'COSMOS' group, which certifies the ingredients are made from at least 95 per cent natural ingredients. If it is labelled

'COSMOS Organic', 95 per cent of the ingredients are organically grown. I also look for products with the 'Leaping Bunny' certification, which guarantees that the company does not test their products or ingredients on animals.

TEXTILES

The fashion industry can be as corrupt and destructive as the fishing industry and shares a similar lack of credibility when it comes to certifications. Some certifications for textiles are more typically found on company websites rather than sewn into products. The ones I see most often are Global Organic Textile Standard (GOTS), Remake Approved Sustainable Brand and Eco-Stylist. These certifications can be ambiguous and difficult to validate as there are so many layers within the production of a garment. When focusing on just the fibre, you are looking at how the plant that created the fibre for the textile was grown. Some plants are more environmentally impactful than others. If the product is made from a synthetic fabric, it's important to remember that it has been derived from a fossil fuel. Finally, if the product is made from recycled fibres, look at what the percentage is.

ELECTRICAL APPLIANCES

Yes, electrical appliances require a lot of energy to run. But since most of us are unlikely to stop using our electric kettles or washing machines, we can at least make the most energy-efficient choices possible. The certification you will likely have seen on everything from laptops and refrigerators to televisions and light bulbs is the EU Energy Label. This label rates the product's energy efficiency in a sliding scale from G to A. This label often features either a European Union or a Union Jack flag.

ETHICAL CERTIFICATIONS

The Fair Trade movement that began in the late 1940s grew out of a radical demand for a power shift, challenging how wealthy nations had been appropriating the benefits of the trade relationships. Fair Trade inspired the creation of a range of ethical certifications. Yet, it is important to realize that a lot of certifications make money because of the high demand of brands wanting to be certified, which can, ironically, make the ethics of these certifications difficult to judge. There have been a variety of exposés on certifications questioning their validity, and some farms and manufacturing facilities have been found to fall short of the requirements for the certifications they claim. At the same time, when used with integrity, these certifications can enable farmers, growers and makers to receive a higher premium for what they are making, which is a positive for them and their livelihoods. It can also support fair working conditions, skill training and leadership empowerment. The two I've seen most are Fairtrade International and the Rainforest Alliance.

Be wary when brands use terms such as 'fair trade' or 'ethically made' in their marketing messaging or packaging design without also identifying with a certification or at least verifying what makes their products so. If a brand has true ethical standards, it should provide in-depth information on its website.

o o o

When you come across certain marketing language, a label or a certification, a useful way to try to understand it is to think back to the Big-Picture Buying Guide. As we know, there really is no way to define 'good' or 'bad', or 'right' or 'wrong'. We each have different interpretations of what that means, and third-party certifications and marketing firms do too. What we *can* do is ask questions, do our research, hold brands accountable, reach out to our local representatives to demand change and encourage others to do the same through community organizing, social media or one-to-one conversations with friends and loved ones.

As in every area of your life, seemingly small changes in your shopping habits can eventually make a big impact. Potential for change lies in each of us, in the choices we make every day, and the approaches we have through our lifetime. It takes just 3.5 per cent of a national population to create meaningful change. So, what are some actions we can take? Here are a few of my favourites to get you rolling:

Daily Actions

- ⊙ Start slow. Choose one area of your home to begin making changes in your purchasing habits. Because I was concerned with pollution from single-use plastics, I started focusing on that material first in my kitchen, then bathroom and slowly worked throughout my home.

From left to right: A shopping basket at the start of my journey with 10 per cent of the items free of single-use packaging, then 20 per cent, and then 60 per cent.

- Shorten your supply chain by buying locally and seasonally.

- Know when to compromise. Don't get frustrated if it feels like you are taking a step back in changes you are making. Find joy in the little wins.

- Think twice before grabbing something to go. How about taking a moment to sit or cook for your own health and the planet.

- Make incremental progress. When I started looking at single-use plastics with my food shopping, I began by working to have 10 per cent be free of single-use packaging, then 20 per cent, and so on. Slowly build at a pace that works for you.

Yearly Actions

- Explore how not buying something can be an act of resistance.

- Tend and care for your items. This is explored in more detail in the Go Keep chapter.

- Carve out an hour every month or every other month to send some emails to your elected officials and favourite brands. Let them know your feelings about their policy and initiatives.

- Join a local community focused on an environmental cause you are interested in.

- Think about making or growing your own things.

- Explore local flea markets and local or online secondhand stores.

Lifetime Actions

- Be gentle with yourself as you progress through change.

- View life through a lens of systems thinking rather than isolated events and objects.

- Respect all that goes into a product, and seek more items made of quality materials, made in fair conditions, that will last a lifetime.

- Practise gratitude when contemplating all that Earth gives us.

go cook

RECIPES DESIGNED WITH OUR BODIES

AND THE PLANET IN MIND

I love to cook! It is my greatest form of meditation. I drift off to another world, away from the day's events and concerns, and find new perspective in the quiet acts of chopping and stirring. The idea for this book was actually born out of my desire to make a zine on kitchen tools and recipes I had experimented with while learning about our compost and waste recovery system. So, I am very excited to be at this chapter with you, as the intention behind this book is to offer tangible and creative steps we can take to not only nurture our planet but also nourish ourselves and our community. This chapter is full of recipes that attempt to solve a problem, such as not letting those wilted greens and stems of vegetables go to waste, not giving up on delicious dips because they are only sold in plastic, or making sure meals on the go aren't highly packaged and expensive.

The more technology has developed, the more complex and processed our food has become. Due to globalization and the ease of transport, our taste buds have become accustomed to foods from far away 365 days of the year, not just at a rare visit to a special restaurant. If each of us is able to shorten our food supply chain by eating more locally grown food and/or cooking meals from scratch, we can collectively lower the impact the food industry has on our planet and nurture a more regenerative model in farming and agriculture. I believe that cooking is a way to celebrate the work and knowledge of farmers as well as the abundance of the earth, which provides us with soil, water and sun to feed us. As we develop confidence in simple, clean cooking, I hope we use it as a tool to liberate ourselves from overly processed and packaged food, instead cultivating a more intimate relationship with Earth's fruits.

While shopping for the ingredients in this chapter, here are a few things to keep in mind:

⊙ For all ingredients but especially produce, check to see where the item was grown and/or manufactured. How close is it to you?

⊙ Look up what's in season before you head to the shop. This will help ensure that you're not buying lots of produce imported from faraway countries. It's also a nice way to be inspired to cook with what's at its best right now.

⊙ Seek balance and moderation in your shopping trolley. If you notice that your trolley is filled with heavily processed and packaged snack foods, could you instead buy more produce and raw ingredients with which to make snacks from scratch?

⊙ Instead of buying all your ingredients from a large supermarket chain, support the local economy by seeking out smaller independant shops, co-ops, speciality food shops or buying direct from farmers' markets. These shops sometimes have bulk sections and less heavily packaged items.

⊙ You may notice that you use the same pantry staples over and over again in your kitchen. If you cook like me, these could include beans, flaxseeds, coconut oil, oats and olive oil. Try to buy these ingredients in bulk, either using your own reusable containers in the bulk section or in larger packages so the ratio of product to packaging is higher.

⊙ Practise compromise and balance. Some items might not be available in the type of packaging or grown in the exact way you want. Seek joy in the items you are able to find that do reflect your values.

⊙ When you're putting together your menu plan and shopping list for the week, make an alternate list of back-up items in case the ones you're planning to purchase are not available. For example, you may have iceberg lettuce on your list. If the only iceberg lettuce at the store is packed in an unrecyclable plastic shell that you'd prefer not to buy, maybe romaine or little gem lettuce would work instead. Having thought through these back-up options in advance helps me stay calm and organized in the shop.

BREAKFAST

I am one of those seriously annoying people who is raring to go the moment they open their eyes. I think one reason I love the morning is because everything seems possible – my mood hasn't been dampened by anything or anyone, including myself. But let's be real. I also love the morning because I love breakfast foods and coffee! Breakfast is a meal that, when eaten out of the home, seems to double in packaging and cost. By prepping some breakfast items to be taken on the go and making your morning beverage at home, you can significantly lower your use of packaging and increase your budget for other meals.

Go Bars

4 teaspoons ground
flaxseeds

285g (10oz) rolled oats

150g (5½oz) chopped
dried fruit, such as
apricots or raisins

70g (2½oz) sunflower
seeds

70g (2½oz) pumpkin
seeds

1 teaspoon ground
cinnamon

½ teaspoon ground
ginger

Grated zest of 1 large
orange

1 teaspoon baking
powder

Pinch of sea salt

6 tablespoons maple
syrup

60ml (2fl oz) sunflower
oil or melted coconut oil

125g (4oz) peanut butter
(or any nut or seed
butter)

tip • If I am making these for
the week, I usually make them
at night and leave them to cool
on the baking tray overnight.
I wrap my bar for the day in a
cloth napkin that goes straight
into my handbag, and I store
the rest in an airtight container
on the worktop.

I was close to calling these 'cookies' because they taste so
good, but 'bars' sounded a little more suited to breakfast
(even though they are actually round). I promise they are
nutritious as well as delicious! Our time in the morning
can be limited. By pre-baking a batch of these bars, you
can be fuelled up for your week ahead and save yourself
from buying highly processed, excessively packaged and
overpriced granola bars or breakfast foods.

Preheat the oven to 190°C/Gas 5.

In a small bowl, mix together the flaxseeds with 120ml
(4fl oz) water; set aside. This will act as a binder for the bars.
In a large bowl, combine all the dry ingredients: oats, dried
fruit, sunflower seeds, pumpkin seeds, cinnamon, ginger,
orange zest, baking powder and salt. Stir them together
with a wooden spoon and then stir in the maple syrup, oil,
peanut butter and the now-thickened flaxseed mixture. Mix
thoroughly so everything is combined evenly.

Using your hands, divide the dough into seven balls and
shape them into flat disks, roughly 10cm (4in) in diameter
and 2.5cm (1in) thick (these don't really change in size as
they bake, so make them the shape and size you'd like). I like
to use a reusable silicone baking mat to line a large baking
tray. If you don't have a silicone mat, you can grease the
baking tray with sunflower or coconut oil.

Bake for 20–25 minutes or until golden and leave them to
cool before enjoying. •

Jar of Bircher Muesli

45g (1½oz) rolled oats

½ apple, not peeled, grated

120ml (4fl oz) any type of milk or yoghurt

60ml (2fl oz) fresh orange juice

Optional additions: sunflower seeds, chia seeds, dried fruit, cacao powder

1 tablespoon nut butter (if you are using a clean jar)

tip • I love the flavour the apple and orange bring, but you could also mix this up with other fruit in season. Fresh orange juice is naturally plastic free!

The idea for this recipe happened totally by accident. I was rushing to leave the house for a shoot and was trying to put together my favourite cereal for an on-the-go breakfast. All of my reusable containers were dirty, so I poured all ingredients in a nearly empty jar of peanut butter and tossed it in my handbag. By the time I got to the shoot, the flavours in the cereal had naturally mixed beautifully, including getting every last bit of peanut butter off the walls of the jar. This made cleaning the peanut butter jar (which is usually a bit of a nightmare!) so much easier, and it also made my favourite cereal that much better. It can be made in any jar, but when using a nearly empty nut butter jar, it really depends on how patient or how hungry you are to gauge when it's 'almost empty' enough for this recipe.

Add the ingredients to the jar and stir to combine. Screw on the lid, give it a little shake and leave overnight in the fridge to soak. In the morning, give the jar another little shake, and it's ready to eat. •

PLANET CAFFEINE:
How to Brew Coffee and Tea with the Planet in Mind

I will admit, I have strayed from my English roots and prefer coffee first thing in the morning, choosing tea for either a late morning or afternoon boost. There are many different ways to make these simple drinks, each with their own questions to weigh up. Just as each of our morning rituals can differ, so can the environmental impact of where the beans and leaves come from and the methods by which we brew them. The charts below compare different brewing methods and their respective environmental impacts. It is also important to consider the soil the coffee or tea is grown in and the livelihoods of the farmers. The global demand for these beverages is huge, and there are lots of large-scale farms to keep up with the demand, some of which use degenerative farming methods and don't comply with human rights policies when it comes to how they treat their workers. If you are able to, seek out brands that are organic (no harmful pesticides have been used to grow the coffee or tea) and comply with fair labour certifications.

COFFEE

BREW STYLE	ENERGY/MATERIALS CONSUMPTION	WAYS TO IMPROVE
French press	Creates the least amount of waste, as there is no filter. Coffee beans are often grown in one location, roasted in another and finally shipped to your supermarket or coffee shop.	Compost your grounds or use them for an exfoliating scrub (see page 208). Boil the water using an electric kettle, which wastes less energy. Buy beans directly from a local roasting company to reduce CO_2 emissions from transport.
Pour-over	If you use a reusable filter (metal or cloth work), this is a relatively low-waste option.	Compost paper filters and grounds, or better yet, make your own reusable filter (see page 216). Boil your water using an electric kettle.
Drip machines	Similar to pour-over, waste produced depends on the filter. Coffee beans are often packaged in some kind of plastic.	Find or make a reusable cloth filter that works in your machine. Where possible, buy beans in bulk.
Percolators	These don't create any waste in the form of a filter, but as they are usually boiled on the stove, they require more energy than an electric kettle.	Consider switching to a French press. Compost your grounds.
Pod coffee makers	One of the most wasteful ways to make coffee, as the plastic pods are not recyclable with your curbside pick-up.	Consider switching to another machine. Buy a reusable pod. Some companies do offer to take back used pods and recycle them, but a lot of energy and materials are used to make and recycle these.

BREW STYLE	ENERGY/MATERIALS CONSUMPTION	WAYS TO IMPROVE
Bottled iced coffee	Impact depends on the packaging material; plastic is not ideal.	Make your own cold brew! If buying pre-made, look for larger bulk sizes to reduce packaging-to-product ratio, and prefer glass bottles.

TEA

BREW STYLE	ENERGY/MATERIALS CONSUMPTION	WAYS TO IMPROVE
Loose-leaf	This is the ideal method of brewing tea, as you only use the amount of tea leaves you require.	Make your own reusable tea bag (see page 218) or use a metal strainer. Compost your tea leaves after brewing.
Disposable tea bag	There is a broad spectrum of waste impact here. Tea bags can be either made from 100 per cent paper or a mix of paper and plastic. Then there is also the string that can be attached with a staple and label tab.	Look for the least amount of plastic packaging. If you can find them, buy tea bags that are 100 per cent paper and packed loose in a box with no plastic seal. Compost the tea and bag, making sure to remove any staples first.
Bottled iced tea	This depends largely on the material it is packaged in.	Look for larger bottles rather than individually portioned ones. Avoid plastic bottles. Make your own iced tea.

PLANT-BASED MILK

Okay, so now that you are caffeinated, you might like to add a swirl of milk to your coffee. I personally don't love nut milks for a few reasons: nuts use a lot of water to grow, usually in low-rainfall areas where the agriculture is taking water from the system; nuts are usually expensive; and nut milks require a long soaking process to make.

Hemp Milk

MAKES 960ML (32FL OZ)

30g (1oz) hemp seeds

Tiny pinch of salt

1 teaspoon maple syrup or honey, or ½ date, finely chopped (optional)

This was the first non-dairy milk I ever made because I was attracted to how simple and quick it was. Hemp seeds derive from the *Cannabis sativa* plant. This highly versatile plant can also be used for clothing, cooking oil and improving soil health as a cover crop. Hemp is also a complete protein because of its essential amino acids, and it's high in fibre and omega-3 fatty acids.

Place all your ingredients in a high-speed blender, add 960ml (32fl oz) water and blend for 1 minute. Taste and see if you'd like to add more sweetener or salt. If you're using a date as your sweetener, make sure it has been fully blended. Decant the milk into an airtight container (I usually use an upcycled jar with a lid), and store in the fridge for up to 4 days. •

From left to right: Oat Milk (page 126) and Hemp Milk

Oat Milk

95g (3½oz) rolled oats

1 teaspoon maple syrup, honey, finely chopped date or other sweetener (optional)

1 teaspoon ground spice, e.g. cardamom, cinnamon or turmeric (optional)

tip • The pulp that you strain out can be added to the Go Bars recipe (see page 119). If you are not going to use it immediately, store in the freezer.

Like other plant-based milks, oat milk exploded onto the market in recent years. Suddenly, there are so many different brands on the shelves it's overwhelming. What is also overwhelming is looking at the ingredients lists and seeing all of these additives that you can't pronounce. The beauty of homemade plant-based milks is that you can keep them clean and simple. Oats are a very versatile and inexpensive staple I pretty much always have stocked in my pantry.

Put all the ingredients in a high-speed blender, add 960ml (32fl oz) water and blend for 30–45 seconds. The trick with oat milk is to not overblend it, because it may develop a slimy texture. Strain the milk through either muslin or a nut milk bag. A fine-mesh strainer also works; it will still let bits through, but if you strain the oat milk twice, you should end up with a relatively smooth result. Decant it into an airtight container (I usually use an upcycled jar with a lid), and store in the fridge for up to 5 days. •

STALE BREAD SOLUTIONS

I love gluten, so bread hardly ever goes to waste in my house, but bread is actually one of the most wasted items in the average household. So, before you throw away the end of a loaf, is there an opportunity to use its staleness to your advantage? Below are three of my favourite ways to use up stale bread. If bread is something you buy a lot, as you work to use up every part of the ingredients you have with these recipes, you may also start thinking about the packaging the bread came in. If it's something unrecyclable such as soft plastic, could you seek an alternative? That might be baking it yourself, requesting to use your own bag or a brown paper bag at the bakery section of your supermarket, or supporting a small local bakery which might have better packaging options. If you have bought a loaf that you can't possibly finish while it's fresh, you can slice and freeze any extra.

Herbed Croutons

MAKES ABOUT 250G (9OZ)

½ loaf bread, such as ciabatta, diced into 2.5cm (1in) cubes

A few pinches of 3 different dried herbs or spices of your choice (I like to use cumin seeds, paprika, and dried oregano)

Salt and pepper

60ml (2fl oz) olive oil

Transforming your stale bread into croutons is perhaps the simplest and quickest way to prevent it from going to waste. These are great to add to a soup or salad. I don't include exact measurements here because it's really meant as a way to use up whatever leftover bread you have. Just be sure to add spices, herbs and salt and pepper to taste, and use enough olive oil to fully coat each crouton.

Preheat the oven to 180°C/Gas 4.

Place the cubed bread in a large bowl and toss with your chosen herbs and spices, salt and pepper. Add the olive oil and mix with your hands or a spoon, making sure each crouton is oiled and seasoned. Spread the croutons out on a large baking tray and bake for 12 minutes, turning halfway through. If you're not using them immediately or have some left over, store in an airtight container at room temperature and use within 2–3 weeks. •

Golden Milk French Toast

1 teaspoon ground cinnamon

½ teaspoon ground cardamom

½ teaspoon ground ginger

½ teaspoon ground turmeric

Pinch of pepper

240ml (8fl oz) any milk

2 large eggs or a vegan flax 'egg' (2 tablespoons ground flaxseeds mixed with 5 tablespoons water)

Butter or any type of oil, for the pan

4 thick-cut slices stale bread (a sourdough type works best, but use up whatever you have)

Seasonal fruit and maple syrup, to serve

I used to think French toast was only for special occasions such as birthdays and had never actually made it myself. But one day I made some delicious golden milk waffles, and it occurred to me the incredible spiced milk from that recipe would be an amazing way to rejuvenate stale bread. This makes the most vibrant, golden breakfast, so I don't wait for birthdays or special brunches to make it. The mix makes enough for two, but if I am making this just for myself (which I definitely do!), I save the golden milk in an airtight container in the fridge to use within two days.

In a medium bowl, combine your spices then your milk and whisk vigorously until the spices have dissolved into the milk as much as possible. Crack your eggs (or add your flax 'egg') into a wide bowl or deep plate that is big enough to hold your slices of bread. Beat the eggs with a whisk, then add your golden milk mix and stir to combine. Heat a couple of tablespoons of butter or oil in a cast-iron or other heavy-based pan over a medium heat. Once the pan is hot, dip the bread into the golden milk mix; try not to oversoak, as it may get a little soggy. You can always squeeze the excess off gently if it does end up sitting a little too long. Fry the bread until golden brown on both sides, 2–3 minutes per side. Serve with fresh seasonal fruit and maple syrup. •

Panzanella Salad

2 large, very ripe tomatoes, roughly chopped

¼ red onion, thinly sliced

½ cucumber, cut into 2.5cm (1in) cubes

Stale bread (1 or 2 slices, or whatever you have), toasted and torn into pieces

A handful of fresh basil, roughly chopped

A handful of flat-leaf parsley, roughly chopped

Salt and pepper

60ml (2fl oz) olive oil

2 tablespoons red wine vinegar

Call me strange, but I love soggy bread. Maybe it's because I love beans on toast, particularly when the beans have had some time to soak into the toast. A panzanella salad is a much more refined way of soaking up sauce with bread, where the sauce is the juice from ripe tomatoes. This dish always seems to taste best when made ahead, so try to prepare it the day before or at least a few hours ahead of serving, to allow all the flavours to come together.

Combine the tomatoes, onion and cucumber in a large bowl (this can be your serving bowl). Add the torn bread, basil and parsley, and season generously with salt and pepper. In a separate small bowl, combine the olive oil and vinegar. Pour the dressing over the salad and mix thoroughly. I like to use the back of a fork to really press the tomatoes so their juice becomes part of the dressing; this helps the bread soak it up. Leave it to sit in the fridge for at least a couple of hours and up to 1 day before serving. •

Reviving Bread

If none of these stale bread recipes hits the spot and all you really want is for your stale bread to be fresh and delicious again, here is a little tip. This only works with unsliced bread and preferably a loaf with a relatively thick crust.

Wet the bread by running the loaf under the tap. Sounds wild, I know, but you are adding back the moisture that has escaped into the atmosphere and made it stale. Then cook in the oven at 150°C/Gas 2 for 10 minutes. Voilà – almost fresh bread, ready to slice and toast. This completes my ode to gluten. •

MAIN MEALS

By now you know I love to cook, but some weeks I just don't have the time to prepare fresh meals for myself. That's why I developed the recipes in this section, which can be batch cooked, stored and eaten throughout the week. I find that once I'm making a mess in the kitchen, it's often worthwhile to make a bit more of something or make two dishes so I'm set up for my week with healthy, satisfying, resourceful meals.

MAKES 4 BURGERS

Earth Burgers

400g (14oz) tin of beans (I love to use black beans, but any bean works), drained and rinsed

270g (9½oz) grated raw root vegetables, such as beetroots, carrots, parsnips, turnips, winter squash or sweet potatoes

120ml (4fl oz) tahini

5g (¼oz) chopped flat-leaf parsley

1 teaspoon ground cumin

1 teaspoon ground paprika

Salt and pepper

2 tablespoons olive oil

tip • This is a great recipe to make then freeze, as the patties are perfectly portioned. I have found that it's better to cook them first rather than freeze them raw. This way they retain more flavour and hold together much better.

I have never actually eaten a 'real' burger, as in one made of beef. My ideal burger is one packed full of whole vegetables that are as unprocessed as possible. There are many new plant-based meat substitutes that claim to emulate the texture and taste of meat, but I personally want my burger to taste and look like the delicious vegetables it's made from.

I used to rely heavily on frozen veggie burgers for those busy nights. But most shop-bought options are wrapped in some type of plastic packet that I couldn't upcycle or recycle. By cooking veggie burgers from scratch, you can avoid that packaging and also shorten your supply chain. I finally got around to making my own burgers, and, after perfecting my recipe, it's hard to go back to shop-bought ones.

Pour the beans into a large bowl and use the back of a fork to mash them slightly (I personally like to keep some of the beans intact, so I only mash about 60 per cent). Add the grated vegetables, tahini, parsley, cumin, paprika and salt and pepper to taste. Mix everything together with your hands. I like to use my hands as it combines everything evenly, and it's nice to touch something other than a keyboard or phone!

Now divide the mixture into four portions, scoop them out one by one, and use your hands to mould them into patties. The trick here is to try to push the mixture together

// continued

gently but firmly, aiming to make the patties as compact as possible. Making them compact will help them stay together when you cook them. I use the inside edge of my thumbs to help give the burgers a hard edge, rather than have loose vegetables popping out. Another tip is to not make the burgers too thick or they will take too long to cook through; about 2cm (¾in) is ideal. I have also made six smaller burgers with this recipe, which works great, too!

When you have shaped your patties, heat the olive oil in a frying pan over a medium-high heat. Add the burgers, in batches if necessary, and cook for 6 minutes per side. (You can also cook them on your outdoor grill.) Serve alone or with your favourite burger fixings. •

Rustic Spaghetti with Two Sauces

200g (17oz) 00 flour or plain flour

1 large egg

90ml (3fl oz) water

50g (1¾oz) semolina flour

tip • When I am cooking for just myself, I usually make enough for two servings and reheat it or work it into a salad the next day. It's best stored cooked, not raw, in an airtight container in the fridge.

I know I say I love a lot of things, but I love pasta. I feel nervous when I don't have all the ingredients for a pasta dinner in my kitchen as it is my standby comfort food and go-to meal when I haven't had time to food shop. I have lost count of how many bowls of pasta have fuelled the writing of this book. Making pasta from scratch does take up a lot of time, but you probably already have the ingredients. I have opted for an egg pasta, but you can make it without.

This is a very rustically rolled spaghetti that doesn't need a pasta press, just your patient hands. The beauty of this recipe is that the shapes don't have to be perfect or too even, but you do generally want them approximately the same thickness so they will cook evenly. Even my friends' kids have made these rustic spaghetti noodles. It's a fun and tactile recipe!

Clean a surface in your kitchen or use a large wooden board. Pour the flour onto your surface like a mini mountain, then make a well in the middle like a crater. Crack the egg into the crater, then whisk with a fork, using gentle movements. Once the egg is mixed, begin to slowly, little by little, incorporate the flour from the edges, until the egg is fully absorbed. Now add the water, little by little, mixing it in with either your fingers or a fork. Continue to work the water into the flour until the dough begins to come together. At first it will be somewhat crumbly, but the more water you add, the more the dough will bond together. As it's taking shape, make sure to collect any excess flour on the surface. Now the dough should be ready to start kneading.

I like to have some extra water and flour handy in case the dough feels too wet or too dry as I work with it. Knead for 8–10 minutes, pushing the dough forwards and backwards against the surface. When the dough feels silky and smooth to the touch, it's ready. Shape the dough into a

// continued

long baguette-like shape about 7.5cm (3in) wide. Now place the dough in a lidded container or a bowl covered with a tea towel. Leave it to sit at room temperature for 30 minutes. You can also leave this in the fridge overnight, if you are prepping ahead.

When you're ready to roll, put the semolina flour in a wide shallow bowl or on a deep plate, and make sure you still have that handy glass of water nearby.

Flour your surface lightly, and roll the dough out with a rolling pin or wine bottle into a thinner baguette shape, about 5mm (¼in) thick and 12.5cm (5in) wide. With a knife, cut off a 1cm (½in) strip. Cover the remaining dough with a slightly damp cloth while you work with this first piece.

Place your fingers on the strip of dough and, keeping them straight, slowly apply pressure down and outwards while rolling. Roll out to the ends of the strip; keep repeating this action and you will be left with a lovely, rustic piece of spaghetti! If your hands get too dry while you're working, dip them in a little water; if the dough sticks to the surface, flour it lightly. It takes practice to find the right pressure and technique to it. A balance of rolling down and out is key, so your hands begin to part as the spaghetti gets longer. Keep cutting 1cm (½in) strips from the remaining dough and repeating the action. When you finish each piece of spaghetti, put it in the bowl of semolina.

Cook the spaghetti in a big pan of salted boiling water for about 4 minutes, drain and enjoy with either of the following sauces. •

// continued

Zesty Bean Sauce

5 tablespoons olive oil

70g (2½oz) walnuts, roughly chopped

½ teaspoon red pepper or dried chilli flakes

2 garlic cloves, sliced

1 sprig rosemary, leaves removed and chopped

400g (14oz) tin cannellini beans or any white bean, drained and rinsed

200g (7oz) rustic spaghetti (see page 135) or dried pasta (approx 225g/8oz)

Fresh basil or flat-leaf parsley, roughly chopped (optional)

Zest and juice of 1 lemon

Salt and pepper

I have always been a tomato-pasta girl, but as my love of beans has developed in recent years, I started to experiment with bean sauce for pasta. Beans' starch content gives the sauce a naturally creamy texture. What's especially great about this one is that most of the ingredients can be found in the pantry rather than needing to be fresh.

First things first, get a pan of salted water boiling. Heat 2 tablespoons of the olive oil in a sauté pan. Add the walnuts and red pepper flakes to the pan and cook over a medium-low heat for 2–3 minutes. Transfer to a bowl and set aside (this becomes the garnish at the end).

In the same pan you used to toast the walnuts, heat another 2 tablespoons of olive oil over a medium-low heat and add the garlic and rosemary. Once the garlic is beginning to brown, add the beans and cook for 4 minutes, stirring occasionally. Add the spaghetti to the boiling water (if using dried pasta, start cooking the pasta a little earlier in the process). Before your pasta is finished cooking, steal ½ cup of the cooking water (about 120ml/4fl oz) and add it to the pan with the beans. Mash half the beans with a fork or a wooden spoon to thicken the sauce. Then add the final 1 tablespoon olive oil, the herbs and the lemon zest and juice. When the pasta is cooked, add another ½ cup of the starchy cooking water to the sauce, then drain the pasta and add it to the pan with the sauce. Bring the mixture to a gentle boil. Season to taste with salt and pepper, mix thoroughly, and serve with a sprinkling of the spicy walnuts on top. •

Veggie Tomato Sauce

2 tablespoons of olive oil

1 garlic clove, minced

1 teaspoon dried oregano

1 large onion, finely chopped

1 pepper, any colour (I usually go for red), finely chopped

1 courgette, finely chopped

1 carrot, finely chopped

1 medium broccoli crown, finely chopped

2 tablespoons tomato purée

400g (14oz) tin peeled whole tomatoes

Salt and pepper

This sauce works beautifully with pasta but also lots of other things, such as with some cracked eggs, like a shakshuka. When I make a batch of this and I don't use it all, I freeze the leftovers in an upcycled container or reusable silicone bag for future meals. Feel free to play with the vegetables according to what you have and like.

Heat 2 tablespoons of olive oil in a large, deep saucepan or Dutch oven over medium heat. Add the garlic and oregano and sauté for 1 minute, then add all the chopped vegetables, stir and season with salt and pepper. Cover the pan and cook, stirring regularly, for 15–20 minutes, until the vegetables are beginning to soften. Add the tomato purée and sauté for a minute or so. Now pour in the tin of tomatoes and use your spoon to break them up slightly. To make sure you get every last ounce of tomato out of the tin, fill it up halfway with water, swirl it around and add the water to the pan.

Simmer the sauce over a medium-low heat, stirring every so often, for 25 minutes. Season to taste with more salt and pepper, and serve over pasta or polenta, with bread, or eat straight out of the pan with a spoon. •

tip • A trick I love to do when getting all the tomatoes out of the can is switch the water to red wine (just fill a quarter of the can), swirl it around, and pour the wine into the pan. I call this the 'Bonnie trick' and not because I thought of it but because it's a trick I learned from the only other Bonnie I know. This also works great when you have a glass jar of premade sauce: screw on the lid with the wine inside and give it a good shake. The wine will add depth to the sauce and waste none of the tomato.

Farmer's Pie

1.4kg (3lb) potatoes, peeled and cut into 5cm (2in) pieces

180ml (6fl oz) any milk

5 tablespoons dairy or vegan butter

1 tablespoon grainy mustard (optional)

Salt and pepper

6 tablespoons olive oil

1 medium onion, finely chopped

3 garlic cloves, minced

3 cups diced seasonal vegetables of your choice, such as carrot, fennel and parsnip

1 tablespoon each of fresh rosemary, thyme and sage, plus more for garnish

3 tablespoons tomato purée

170g (6oz) dried green lentils

820ml (28fl oz) veggie stock

75g (2¾oz) roughly chopped seasonal greens, such as Swiss chard or collard greens

tip • This is a great meal to freeze in individual portions for those days you have no time but need all the comfort.

When I was a child, I always used to be so envious watching other kids eat shepherd's pie, because I love mashed potatoes! It took me a while to understand that 'shepherd' meant there was lamb in it (which I didn't grow up eating and still don't). It just looked like a creamy cloud of mashed potato. So I satisfied my dreams of a cloudlike-mash pie and made a vegetarian version to celebrate and honour the farmer, whom we'd be nothing without. I have suggested some of my favourite vegetables to use, but it's extremely versatile, so look into what is seasonally grown in your local area and be inspired by what is available.

I normally cook this in a medium Dutch oven, sautéing the veg and baking the whole dish in the same pot. If you don't have a suitable pot that can go from the stove to the oven, sauté your veg in a sauté pan on the hob and transfer it into a baking dish before topping with mashed potatoes and popping in the oven.

I hope you enjoy this meal as a celebration of the elements that grew the food – soil, water, sun – and those who tended and farmed it, and lastly you, for taking the time to cook it for yourself.

Preheat the oven to 180°C/Gas 4.

Place the potatoes in a large pan with salted water. Bring to the boil and cook until soft, 20–30 minutes. I use a fork to test for doneness. Drain the potatoes, return them to the pot and add the milk, butter, mustard and salt and pepper to taste. Mash until fluffy and smooth.

Meanwhile, while the potatoes are cooking, heat 5 tablespoons of the olive oil in a medium Dutch oven over a medium-high heat. Add the onions and garlic and sauté for about 5 minutes, or until the onions have started to soften. Add the diced vegetables (not including the leafy greens), then add the herbs and season to taste with salt and pepper. Cook for 15–20 minutes, stirring occasionally.

Add the tomato purée and sauté for 1 minute, then add the lentils, stirring to make sure they get coated in all the

flavours. Add the veggie stock, bring to the boil, then reduce the heat to maintain a simmer, cover and cook for 25 minutes. Now it's time to add your leafy greens. Stir them in, cover and cook for another 5 minutes.

Turn off the heat. Top the veggie mixture with the mashed potatoes, using a spatula to smooth out the top. Drizzle with the remaining 1 tablespoon olive oil, garnish with some more herbs, if you like, and bake for 10 minutes, or until the liquid starts bubbling at the edges. If you want a crispy finish, pop under the grill for 5 minutes. •

Rainbow Roast

About 1.8kg (4lb) assorted vegetables, cleaned and chopped (roughly 2 x 2.5cm/1in pieces for veggies in the first two columns and 3 x 2.5cm/1in pieces for veggies in the third and fourth columns)

60ml (2fl oz) olive oil

3 garlic cloves, minced

Salt and pepper

A handful of fresh herbs (earthy ones such as rosemary, sage and thyme work well)

1 teaspoon cumin seeds (optional)

1 to 2 teaspoons red pepper flakes or paprika (optional)

Similar to the Farmer's Pie (see page 142), this dish can be a celebration of what is in season. I love this recipe in autumn and winter months, as I find the seasonal root vegetables very grounding. I like to roast a big tray of vegetables on a Sunday or whichever day is my quietest, so that I can eat them throughout the week in salads, breakfasts or those mid-to-late-week dinners when the fridge is emptying out. I am predominantly led by colours when I am choosing vegetables at the supermarket, but in the back of my mind, I am also balancing out my flavours and greens-to-carbohydrates ratio. I like to build my roasting tray using this table:

CHOOSE ONE		CHOOSE TWO		CHOOSE ONE		CHOOSE ONE
Sweet potato	+	Carrots	+	Brussels sprouts	+	Onion
Purple potato		Parsnips				Spring onions
Potato		Beetroots		Broccoli		Leek
		Swede		Cabbage		
		Squash		Cauliflower		

tip • The more you make this dish, the more you will adjust for how some vegetables cook more quickly than others. For the longer-to-cook vegetables such as potatoes, you want to cut them smaller than, say, a piece of broccoli, so they will be done at the same time.

Preheat the oven to 200°C/Gas 6.

In a bowl, toss the vegetables with the remaining ingredients, mixing with your hands so they are well coated in oil and herbs. Transfer to a baking tray or large roasting tray and roast for 30–40 minutes, turning them halfway through and checking to make sure they don't get overcooked. Eat the rainbow of veggies as they are or pair with a salad, crumble feta cheese on top or add them to a breakfast bowl. Store any leftovers in an airtight container in the fridge for up to 6 days. •

SNACKS

I love snacks. I love crisps, I love dips, I love anything bite-size that can be eaten at any time of day. However, most snacks come highly packaged, heavily processed and overpriced. For a long time, I went without dips at home because I was more concerned with my single-use plastic waste than my desire to eat them. Instead, I just enjoyed them when I would eat out at restaurants. But then 2020 came and eating out at restaurants was less of a thing, so I tried my hand at making my own dips. I must say, I'm not sure what took me so long; they are delicious, fresh and made to your preference.

MAKES ABOUT 225G (8OZ)

Bonnie's Beetroot Dip

2 medium beetroots, washed and cut in half

400g (14oz) tin white beans (of course I chose my favourite, cannellini beans), drained and rinsed

A generous handful of dill

35g (1¼oz) walnuts or sunflower seeds

60ml (2fl oz) olive oil

1 garlic clove

Juice of 1 lemon

Salt and pepper to taste

This vibrant beetroot dip has become a favourite, brightening any happy-hour table, salad or sandwich. I have shared this recipe with many friends, and now you!

Bring a medium pan of salted water to the boil. Add the beetroot and cook over a medium heat for about 30 minutes, or until tender. When the beetroots are cooked, transfer them to a food processor and add all the remaining ingredients. Blend for a minute or two. I don't overblend it as it's nice to keep some of the texture of the beetroot and walnuts. Transfer the dip to a bowl and chill in the fridge for at least 30 minutes, then enjoy with homemade crisps (see page 152) or sliced raw veggies. Store leftovers in an airtight container in the fridge for up to 5 days. •

From top to bottom: Curried Carrot Dip (page 149), Muhammara Dip (page 148), Flatbread (page 151), and Bonnie's Beetroot Dip

Muhammara Dip

1 garlic head

1 red onion, cut into large wedges

2 red peppers, cut into quarters

2 plum tomatoes, roughly chopped

1 small red or green chilli, or 1 teaspoon dried chilli flakes

3 tablespoons olive oil

Salt

1 teaspoon smoked paprika

1 teaspoon ground cumin

A handful of coriander leaves

35g (1¼oz) walnuts or sunflower seeds

2 tablespoons pomegranate molasses, or 1 tablespoon sugar

Juice of 1 lemon

Muhammara is of Syrian origin and is eaten across Palestine and Lebanon. It has a wonderful balance of sweet and spicy. The sweet comes from baking the vegetables till they start to caramelize and the unique sweetness of the pomegranate molasses. A fresh chilli brings the spice; feel free to turn it up or down for your taste. (If you're feeling lazy, which I often am, or can't track down pomegranate molasses, you can substitute sugar.)

Preheat the oven to 225°C/Gas 7.

Lay the head of a garlic on its side and cut it across the equator so you have two halves. I love the pattern of the garlic this reveals! Place the garlic, onion, peppers, tomatoes and whole chilli in a large roasting pan, drizzle with 2 tablespoons of the olive oil, and season with salt, paprika and cumin. Cook for 35 minutes, turning the vegetables halfway through. You'll know they are done when they start to brown and caramelize. Remove them from the oven.

Once the vegetables have cooled slightly, remove the garlic cloves from their skins; they should pop out easily. If you don't want the dip to be super garlicky, you could save half of the roasted garlic for another dish. Stem the chilli and add it to a blender with the rest of the roasted vegetables. Then add the coriander, walnuts, remaining 1 tablespoon of olive oil, pomegranate molasses, lemon juice and a little more salt. Blend to a slightly chunky consistency, smooth enough that you can easily dip in your bread, crisps or crudités. •

Curried Carrot Dip

120ml (4fl oz) olive oil

3 carrots, roughly
chopped

2 teaspoons curry
powder

1 garlic clove, minced

400g (14oz) tin lentils,
drained and rinsed

1 tablespoon roughly
chopped fresh ginger

A handful of coriander

Juice of 1 lemon

Salt and pepper

tip • If the carrots you used
had green tops, don't compost
them, as they make great
pesto! This also works for any
type of loose herb or greens
you have.

Last in our party of dips is another bright and colourful
recipe. I absolutely love carrots. It could be an affinity with
the colour, or I still secretly believe that by eating them I can
see better in the dark.

Heat 60ml (2fl oz) of the olive oil in a medium sauté pan or
Dutch oven over a medium heat. Add the carrots and curry
powder and cook for 15–20 minutes, adding the garlic in
the final 5 minutes. The carrots should be starting to brown
and feel tender when poked with the tip of a knife. Transfer
to a blender or food processor and add the remaining 60ml
(2fl oz) olive oil, lentils, ginger, coriander, lemon juice and
salt and pepper to taste. Blend until smooth and serve. This
can be stored in an airtight container in the fridge for up to
4 days. •

Flatbread

130g (4½oz) plain flour,
plus more to knead

½ teaspoon baking
powder

Pinch of sugar

Pinch of salt

120ml (4fl oz) Greek
yoghurt

Olive oil

A dip is not a dip unless you have something to dip in it. These flatbreads are a perfect accompaniment to enjoy with your dips. They are also super simple and quick to make. I would recommend using Greek yoghurt because you want that sour, acidic taste, but if you can't find it or prefer not to use it, just add a squeeze of lemon juice to whatever yoghurt you have. This is crucial, as you need the acidity to react with the baking soda and relax the gluten to make the flatbreads extra fluffy and soft. Enjoy these flatbreads dipped in one of my dips, or use them to soak up the juices of any dish you are cooking.

In a medium bowl, mix together the flour, baking powder, sugar, salt and yoghurt with a wooden spoon. Once you have a rough, shaggy dough, transfer it to a clean, floured surface. Knead for about 4 minutes, or until the dough feels smooth and almost silky. Cover with a slightly damp tea towel and leave to rest for 20 minutes. Cut the dough in half, shape each half into a ball, and use a rolling pin to roll them into approximately 3mm (⅛in) thick flatbreads (you could also split the dough into quarters and make four smaller ones). Sprinkle a little flour on each side of the flatbreads; this will help them not to stick to the pan.

Heat a large cast-iron pan (or any frying pan) over a medium heat and add a tablespoon or two of olive oil – enough to grease the pan but not so much that you end up frying the bread. Once the oil is hot but not smoking, place your flatbreads in the pan, in batches if necessary, and cook until air starts to puff up inside, flipping them every minute or so, about 5 minutes total. Serve immediately. •

Flatbread and Curried Carrot Dip

Potato Crisps

2 russet or Yukon gold potatoes

240ml (8fl oz) sunflower oil

Sprig of rosemary and a clove of garlic (optional)

Salt

tip • Once you've fried and snacked, you can save the leftover oil in an airtight jar and use it again the next time you want to make crisps.

Quite possibly the best snack in the world. Ask any one of my friends and they will confirm that I have an addiction to crisps, as I was born calling them (or 'potato chips')! Sadly, they come packaged in the near impossible-to-recycle soft plastic or mixed-material packets, yet the raw ingredient of potatoes you can buy loose. A simple solution to avoid the packaging is to fry crisps yourself. It may sound time consuming, but trust me – the crackling of the oil and the taste of fresh potato crisps is so worth it.

Wash the potatoes, towel them dry and cut them into very thin slices, about 3mm (⅛in) thick. You can use a mandoline for this, but I have found a sharp knife works just great. As you cut, lay your potato slices out on a tea towel (use a dark one or one you are not particularly attached to, as the starch can stain bright white cloth). Once they are all sliced and laid out, leave them to dry for a few hours. I have rushed this step before, and there was too much moisture for them to become crisps.

Once the potato slices are dry, heat up the oil in a deep pan over a medium to high heat. Add the rosemary and garlic, if using. Once the oil is crackling hot, remove the rosemary and garlic. Add the potato slices in batches. The amount per batch will vary depending on the size of your pan – they can overlap a little bit, but don't overcrowd the pan. Flipping them occasionally, fry the chips until they start turning a light to dark brown colour. Every potato is different, and cooking time depends on the thickness of your slices, so knowing when they're ready really comes down to experience and taste. I fry my crisps for 3–4 minutes on average. Take out the crisps with a slotted spoon, allowing as much oil as possible to drain before putting them on a plate. Then salt to taste and enjoy! •

From top to bottom: Potato Crisps, Chocolate Brittle (page 155) and Candied Orange Peel (page 154)

Candied Orange Peel

4 large oranges

400g (14oz) sugar

A handful of chamomile flowers (optional)

tip • If you have too many oranges or cut too many peels, you can use them to make a surface cleaner (see page 213). Don't forget to eat the oranges or squeeze them for juice, too!

I wasn't a fan of candied citrus until I found myself filming food processes for a short film I directed. I experimented with different foods and cooking methods to find visually inviting, textural and colourful scenes. I thought that the bubbling bright orange citrus zests would look beautiful, so I found myself making candied orange peels. One additional colour and texture I added was chamomile flowers, from my loose-leaf tea collection. This fun, creative experiment turned out to be a delicious snack I didn't know I loved. These are also delicious dipped in chocolate.

Use a knife or peeler to remove the peels from the oranges. Place them on a cutting board, and carefully cut away as much of the white pith as possible (this will keep your candied peels from being too bitter), then cut the peels into 5mm (¼in) wide strips.

Place the orange peels in a medium saucepan and cover with water. Bring to the boil, then reduce the heat to low and simmer for about 15 minutes. Drain off the water, return the saucepan and peels to the stove and add the sugar, 480ml (16fl oz) water and the chamomile flowers.

Bring to the boil, then reduce the heat to maintain a simmer and cook gently, stirring with a wooden spoon or silicone spatula, for about 1 hour. While the syrup is bubbling away, arrange a wire rack over a baking tray lined with an old tea towel. When the syrup has almost completely reduced, transfer the citrus peels to the rack and use a spatula or wooden spoon to separate each piece (be careful – they will be very hot!). Once they have cooled to room temperature, cover with a clean tea towel and leave to dry overnight. Store in an airtight container in a cool, dry place. Will keep for at least a month. •

Chocolate Brittle

This is a great little trick to transform plain chocolate into fancy chocolate without having to become a chocolatier.

Bring about 5cm (2in) of water to the boil in a saucepan and top with a heatproof bowl over the pan. Break a bar of dark chocolate into the bowl and let it melt over the simmering water. Pour the melted chocolate into a reusable, heat-resistant silicone bag. Throw in any desired fancy ingredient such as chopped almonds, pumpkin seeds, goji berries, freeze-dried fruit or sea salt. The possibilities are endless! Seal the bag, shake it up a little, and place it flat in your freezer for 20 minutes. Once the chocolate hardens and while the bag is sealed, break it up with your hands, then open it up and snack away. Store any leftover chocolate brittle in the fridge for future snacking. •

PLASTIC-FREE PICNIC IDEAS

Back in 2018, I organized the first ever plastic-free picnic with Greenpeace in Long Beach, California, while their ship, the *Arctic Sunrise*, was docked there. I put together a great panel of speakers on single-use plastics and the health of the oceans, and all the guests at the event were invited to bring their own picnic, free of single-use plastic. The great thing about this was that it engaged everyone on a personal level before the event had even begun. Each participant used their imagination to think about what plastic-free food options to bring, and that was an action within itself. People looked excited as they showed up, and it was clear that they already felt connected with the issue we were going to discuss in the panel. The ideas could even spark some great conversation over the picnic blanket. Here are some tips to think of when preparing for a picnic free of plastic.

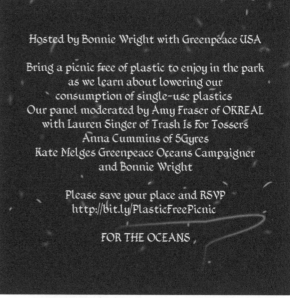

- Pack your snacks in reusable containers.

- Bring your dressings in a separate glass jar so salads can be dressed freshly at the picnic.

- Bring reusable utensils or a set from home. I like to wrap mine in a cloth napkin to protect them and also to use at the picnic instead of paper towels.

- If you are not cooking or preparing your foods at home and are instead buying them on the way to the picnic, here are some of my suggestions that I usually find free of plastic: fresh fruit and vegetables, bread from the bakery section, glass jars or cans of olives or marinated vegetables such as artichokes and roasted red peppers, some bulk items such as nuts and dried fruit. Made-to-order sandwiches can be wrapped in paper or burritos wrapped in foil.

- Bring a flask or large mason jar with a homemade drink, such as my Jamaica Rose Iced Tea (see page 168).

PRESERVES

Preserving fresh food with salt or sugar is a great way to extend the life of some of my favourite ingredients. It's one of the oldest and most reliable methods of food storage we have, and it also yields some seriously delicious flavours!

MAKES ABOUT FOUR 225G (8OZ) JARS

Rustic Cafe

Rustic Café's Apple, Rhubarb and Ginger Jam

3 apples (I use Pink Lady apples)

600g (1lb 4½oz) rhubarb (5 stalks)

1 (roughly 10cm/4in) piece fresh ginger

700g (1lb 9oz) caster sugar

Juice of 3 lemons

tip • To know when the jam has reached its setting point, put a small plate or saucer in the freezer. When you think the jam is ready, and the plate is cold, put a small amount of jam on the plate and gently push it with your finger. If it wrinkles and looks like it has a slightly harder outer layer that prevents the jam from running and filling in where your finger mark is, it's set. If it doesn't, keep cooking the jam, then test again.

I love making jam. It's such a fun, easy way to preserve any fruit at the peak of ripeness and enjoy it throughout the year. By simply cooking down fresh fruit with lots of sugar and a little lemon for acid, you can extend the life of your favourite fruits. My friend Emily Eisen and I started a Jam Club with our pop-up concept Rustic Café when the pandemic prevented us from organizing in-person events. By making jam, we were still able to connect with our community and continue donating all of our proceeds to our friends at SUPRMARKT, an affordable organic grocery serving low-income communities in South Central LA. We make our jam with organic and seasonal fruit from California, every other month, for our Jam Club subscribers. Here is a recipe for our Apple, Rhubarb and Ginger Jam, my favourite flavour we have made so far!

Core and peel the apples and compost the skins. (The worms in my worm compost bin love apple skins!) Dice the apples and cut the rhubarb into 2.5cm (1in) pieces. Peel and grate the ginger with a Microplane. Add these ingredients to a large pan, along with the sugar and lemon juice. Turn the heat to low and cook, stirring, while the sugar slowly dissolves in the juices from the fruit. Once the sugar has completely dissolved, turn up the heat and bring to a rolling boil. Boil until the setting point has been reached, usually 15–20 minutes (see the tip).

Pour the jam into sterilized jars. I like to use upcycled jars and sterilize them either by pouring over boiling water or putting them in a low oven without their lids on. Once you have filled the jars, put the lids on and leave them to cool. Once they are cool, tighten the lids a little more. Refrigerate the jam and consume it within a month. I find this easy to do, as I eat jam and peanut butter on toast multiple times a week. •

Preserved Lemons

MAKES ABOUT 500G (1LB 2OZ)

About 5 lemons (this will depend on the size of your lemons and the size of your jar)

About 60g (2oz) salt

Peppercorns

Extra lemon juice, as needed

I once lucked out with a big box of lemons from a friend whose family had a tree. I couldn't eat them all before they went bad, so I decided to preserve some. You can find preserved lemons in cuisines across the globe, from Morocco to the Middle East, and in Ayurvedic food from India. They're great for brightening and adding depth to dishes like tagines, and are wonderful in salad dressings and drinks.

Quarter the lemons from the top (opposite the stem end), stopping about 2.5cm (1in) from the bottom so the lemon quarters remain attached. Rub salt inside the lemons, without breaking them apart, and stack them into a mason jar, pressing a bit to squeeze them in and release their juice. When you can't fit any more lemons, add the peppercorns and top off with lemon juice, making sure the lemons are totally submerged. Screw the lid on tight and tip the jar upside down a few times to mix everything up. Leave out at room temperature for 2–3 weeks, turning the jar upside down and back once a day. When you want to use them in a dish, rinse them first, then cut out and discard the flesh, as it's the rind you eat. The jar can be stored in a cool and dark place for up to a year. •

Speedy Pickled Vegetables

120ml (4fl oz) rice vinegar or apple cider vinegar

1 tablespoon sugar

2 teaspoons salt

100g (3½oz) thinly sliced vegetables, such as onion, cucumber and carrot

I have never been that keen for pickles; the closest I get to anything pickled is sauerkraut. I think this is because I generally find the brine too intense. These quick-pickled vegetables are milder in flavour than your average pickle, as they are only sitting in the vinegar for 15 minutes before being eaten. They can be made when you need them and can transform a simple meal with a pop of colour and crunch.

In a small bowl, whisk together the vinegar, sugar and salt. Add the vegetables and give them a good stir. If I'm making these for guests, I like to do them in three separate bowls, but if it's just for me, I mix all the veggies in one. Leave the pickles to sit in the brine for 15 minutes, pressing them down from time to time to make sure they are fully submerged. Then enjoy with just about anything! •

CONDIMENTS, OILS AND STOCK

Essentials such as condiments, oils and stock hold a lot of power in the kitchen. Oils and stocks are the building blocks for so many of my favourite dishes, and I can't imagine a world without condiments. A simple meal, with a great condiment, quickly becomes a delicious meal.

Tomato Ketchup

MAKES ABOUT 240ML (8FL OZ)

370g (13oz) tomato purée

3 tablespoons maple syrup

3 tablespoons apple cider vinegar

2 teaspoons onion powder

2 teaspoons dried oregano

1 teaspoon salt

Full disclosure: as a child, I would eat tomato ketchup like it was hummus. I would dip carrots in it! Carrots are still my favourite crudité, but I now use tomato ketchup as a condiment not a dip. This recipe pairs perfectly with my veggie burgers (see page 133).

Mix all the ingredients in a bowl. Store in an airtight container in the fridge for up to a month. •

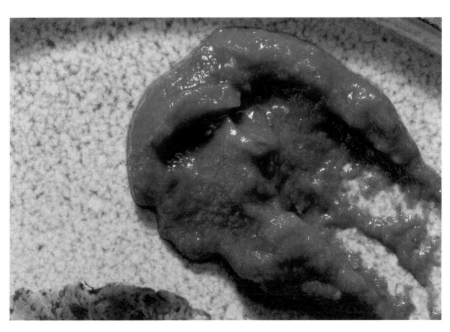

Dressing in a Jar

60ml (2fl oz) olive oil

Juice of ½ lemon, or
1 tablespoon white wine
vinegar

1 garlic clove, crushed
(not cut)

1 tablespoon whole-
grain mustard

1 teaspoon salt

½ teaspoon pepper

1 tablespoon tahini
(optional; if using, add a
little more lemon juice)

This is a very simple salad dressing recipe, but I wanted to
highlight the trick of using an old jam or condiment jar
to both make the dressing in and transport it. There is
nothing worse than taking your beautifully prepared salad
to work or school or a picnic and it's all soggy because it's
been sitting in the dressing for too long. No more soggy
salads here!

Combine all the ingredients in a small upcycled jar with a
lid. Screw the lid on and shake vigorously. If there are clumps
of mustard or tahini, use a fork or spoon to blend them
in. Taste for seasoning, and add more acid, salt or pepper
according to your preference. Put the lid back on and give
it another good shake. Store in the fridge for up to 3 days.
Before dressing your salad, remove the garlic clove. •

Herb-Infused Oils

**4 herb sprigs (I like
thyme and rosemary)**

Olive oil

This is a great way to preserve fresh herbs and jazz up
your olive oil. If you grow your own herbs – a guide to
which you can find on page 226 – this is a perfect way to
use up your harvest.

tip • These also make lovely
gifts!

Once you have picked or bought your herbs, hang them up
to dry. You want them as dry as possible, so depending on
the herbs, this could take anywhere from 2 days to 2 weeks.
Choose a nice bottle to upcycle (I love using an old wine
bottle with a stopper or pourer), add the sprigs of dried herbs,
and top with olive oil. You can use the oil immediately, but
the flavour will become much more pronounced as it sits and
infuses. It will keep for 4–6 weeks. •

Veggie Scrap Stock

MAKES ABOUT 960ML
(32FL OZ)

900g (2lb) assorted
vegetable scraps

1 red or yellow onion,
cut in half

A couple of carrots, if
you have them

A couple of celery stalks,
if you have them

3 garlic cloves, peeled

2 bay leaves

2 tablespoons olive oil

Salt and pepper

tip • I like to use this stock
as the base for a soup. Every
week or so, I clear out my
fridge and make what I call
an odds and ends soup –
throwing in any sad vegetable
bits, leftover grains or beans,
Parmesan cheese rinds, and
fresh or tinned tomatoes.
It looks and tastes a little
different every time, but it
always hits the spot.

I love how food scraps that we might have called 'waste' can make the tastiest meals, and this veggie stock is a perfect example. Instead of tossing my vegetable scraps in the compost bin, I save them up and, when I've collected enough, transform them into a delicious, nourishing stock. You can throw almost anything into this stock – onion ends, root veg, turnip greens, potato and carrot peels, any fresh or dried herbs – but I recommend avoiding artichokes, members of the brassica family, such as cabbage and broccoli, and salad greens. I try to add some fresh onion, carrot and celery when I'm making this, but don't worry if you don't have them. The scraps alone will still make a beautiful, flavourful stock.

Place all of the vegetables and the bay leaves in a large stockpot or casserole dish and cover with cold water. Add the olive oil and a good pinch each of salt and pepper. Bring to the boil over a high heat, then reduce the heat to maintain a simmer and cook for 30 minutes. Strain the stock, discarding the solids into your compost bin, and season to taste with more salt and pepper. Either use immediately or store in the fridge for about a week or the freezer for up to 3 months. •

DRINKS

Drinks such as tea and sodas often come in wasteful packaging, are shipped from far away and contain additives to prolong their shelf life. A simple way to shorten this food supply chain and make drinks as clean as possible is to make them at home.

MAKES 5 SERVINGS

Ginger 'Bug' Soda

1 large piece fresh ginger

1 tablespoon plus
5 teaspoons sugar

960ml (32fl oz) chilled brewed tea (I like hibiscus, chamomile or tulsi), for serving

Sugar or honey, for serving

This is my lazy take on kombucha. The only true fermentation I have the patience for is making a sourdough starter for bread. I have yet to try building a SCOBY for kombucha, but this ginger 'bug' is a quicker and simpler process for making a naturally carbonated drink.

Combine 1 tablespoon grated ginger (you can keep the skin on), 1 tablespoon sugar and 240ml (8fl oz) water in a pint-size mason jar. Stir with a spoon until the sugar is mostly dissolved, then cover with a napkin or any piece of breathable fabric. Leave the mixture to sit and ferment at room temperature for about 5 days, stirring in 1 teaspoon sugar and ½ tablespoon grated ginger each day. When your bug is bubbling and has started to smell yeasty, it's ready.

To serve, strain out the solids and mix the bug with the tea. Add sugar or honey to taste, and serve cold. •

Jamaica Rose Iced Tea

MAKES ABOUT 960ML
(32FL OZ)

50g (1¾oz) sugar

½ cinnamon stick

4 thin slices of fresh ginger

4 allspice berries

40g (1½oz) dried hibiscus flowers

Ice

Sparkling or still water

Lime juice

tip • The hibiscus flowers you are left with after you strain the tea make an incredible taco filling. I came up with my own recipe after trying a super tasty Jamaica Rose taco from a taco truck. To make this, lightly rinse the hibiscus flowers and season them with cumin, paprika, salt and pepper. Then cook them in a sauté pan with some olive oil over a medium heat for about 10 minutes.

Ever since travelling to Guatemala with Rainforest Alliance to learn more about sustainable forestry, I have been obsessed with Jamaica Rose Iced Tea. I was lucky to enjoy this cooling and refreshing beverage in all corners of the country, including beside dormant volcanoes in Lake Atitlán and around a family's table in the rain forest of the Maya Biosphere Reserve. This recipe is a homage to that trip and the incredible stewards of the rain forest I had the pleasure to meet.

Fill a small saucepan with 960ml (32fl oz) water, then add the sugar, cinnamon, ginger and allspice. Bring to the boil and stir until the sugar has dissolved. Remove from the heat and add the dried hibiscus flowers. Cover the pan and leave to steep for 20 minutes. Strain into a jug or mason jar. You are left with a beautiful deep Jamaica Rose concentrate. Best enjoyed over ice: fill a glass halfway with the Jamaica Rose concentrate, then top it off with sparkling or still water and a squeeze of fresh lime juice. *Delicioso!* •

Go Gently Tea Blend

MAKES 1 SERVING

1 tablespoon loose-leaf peppermint tea

1 tablespoon loose-leaf tulsi tea

1 tablespoon loose-leaf chamomile tea

Tea is my go-to comfort drink when I am anxious, run-down, on my period or in need of some soothing. I like to make a big pot with loose-leaf tea using my reusable tea bag (see page 218). In the evenings, I drink it paired with some chocolate brittle (see page 155).

Combine the tea leaves in a reusable tea bag or loose-leaf tea strainer, place in a teapot and cover with hot water. Leave to steep for 5 minutes. Strain into a mug and enjoy! •

SWEET TOOTH

I have bookended this chapter with two of my favourite meal types, breakfasts and sweets! When I was a child, I was convinced that we had two stomachs: one for savoury food and one for sweet food. Needless to say, I never managed to convince my parents of this groundbreaking discovery.

Pantry Choc Chip Cookie

MAKES 12 COOKIES

2 tablespoons ground flaxseeds

90g (3oz) oat flour or plain flour

120g (4oz) coconut sugar or sugar

100g (3½oz) dark chocolate chips (or cut up a chocolate bar with a knife)

75g (2½oz) peanut butter (or any nut or seed butter)

60ml (2fl oz) coconut oil, melted

1 teaspoon vanilla extract

½ teaspoon bicarbonate of soda

½ teaspoon salt

What's so great about this cookie is it requires no fresh ingredients, so it's perfect for satisfying a sudden, unplanned sweet craving. I always have these ingredients on hand in my pantry, and I have a strong feeling that after you make these once, you will be sure to always have them, too!

Preheat the oven to 190°C/Gas 5.

In a small bowl, mix the flaxseeds with 5 tablespoons water to make a flax 'egg'; set aside. In a large bowl, mix together the remaining ingredients. Add the flax 'egg', which will now have a thicker consistency, and stir to combine.

Using your hands, roll the dough into golf ball-size balls and lay them out on a baking tray. Make sure you give them space – they expand as they bake. Flatten each ball of dough slightly with your hands to about 2.5cm (1in) thick. Bake for 10–12 minutes, then remove from the oven and leave to cool for 5 minutes. Trust me, do try and wait; if you remove the cookies from the tray too early, they will likely crumble. •

tip • I usually don't bake all my cookie dough at once. Store unbaked cookie dough balls in an airtight container in the fridge for up to a week and bake a few more when the next craving strikes!

Aquafaba Chocolate Mousse

100g (3½oz) dark chocolate, roughly chopped

120ml (4fl oz) aquafaba (or roughly the liquid from a 400g/14oz tin of chickpeas)

Toppings of choice, such as fresh fruit and chopped nuts

When it comes to dessert, I will always go for the chocolate flavour of anything. In lots of the photos of me as a child, I have the remnants of chocolate around my mouth. This mousse is a great way to impress any dinner guest and, apart from your desired toppings, it requires only two ingredients. But first, what is aquafaba? It is the cloudy water left over from cooking chickpeas, either when you have cooked them yourself or, as I more commonly find it, in a tin of chickpeas.

Melt the chocolate in a glass bowl set over a pan of simmering water. Transfer the melted chocolate to a bowl and put it in the fridge for 10–15 minutes. (If the chocolate is too warm when you add it to the whisked aquafaba, it will deflate it too much.)

While the chocolate is cooling, whisk the aquafaba with an electric mixer for 10–15 minutes. You can also use a blender, a hand mixer with a whisk attachment or a food processor. By whisking the aquafaba at a high speed, you will witness it doubling in size and transforming into a fluffy, cloudlike consistency; it's really quite extraordinary. You can't overwhisk aquafaba like you can with eggs, so keep going until this fluffy consistency happens.

Once the chocolate is slightly cooled (not solid), add it to the aquafaba, stirring it in as you add. The aquafaba will deflate slightly and seem watery, but that is okay. Transfer the mousse to serving bowls and refrigerate for at least 30 minutes or until you are ready to eat it. The mousse will now be firm and can be decorated with any toppings. I love having mine with strawberries or raspberries and crumbled nuts. •

Ice Lollies

Ice lollies are a great way to use up overripe fruit – plus, they're refreshing to eat in the summer months. When you think of overripe fruit as frozen juice lollies or smoothies, the possibilities are endless. In my continuing effort to lower my use of unrecyclable materials, I have enjoyed experimenting with flavours and making my own ice lollies in a metal mould. Here are two of my favourite flavour combinations.

Berry & Banana

85g (3oz) fresh or frozen raspberries, strawberries, blueberries or a mix

2 ripe bananas

120ml (4fl oz) plain yoghurt

Small pinch of salt

2 tablespoons honey (optional)

Combine all the ingredients in a blender and blitz until smooth. Carefully divide the mixture among six ice lolly moulds and freeze until solid, at least 3 hours. •

Lemon & Thyme

100g (3½oz) sugar

3 sprigs thyme

180ml (6fl oz) fresh lemon juice

Combine the sugar, thyme and 120ml (4fl oz) water in a small saucepan. Bring to a simmer and cook for a couple of minutes to dissolve the sugar. Now turn off the heat, cover, and leave the thyme to infuse in the simple syrup for at least 10 minutes. Strain the syrup into a bowl and whisk in 180ml (6fl oz) cold water and the lemon juice. Carefully divide the mixture among six ice lolly moulds and freeze until solid, at least 3 hours. •

go keep

PRACTICES TO TEND AND MEND
YOUR THINGS TO PROLONG THEIR LIFE

It can sometimes be more within our nature to consume and waste than to tend, care for and maintain what we already have. The practices I have learned and discovered are by no means new ideas; my grandparents' generation would have more naturally taken care of what they owned. Out of pride, less access to replacements and less of a culture of consumption, my grandparents would tend and mend their possessions. I didn't think much of it as a child, but my concern for the health of this planet and humanity has led me to reconnect with their long-held practices of preservation.

There are many communities and cultures outside my own that regularly practise these skills and ways of being either out of necessity and ingenuity or because they have successfully passed down the knowledge through generations. I have a deep respect for cultures that revere storytelling and skill-sharing.

As I have explored practices that best help me to nurture my things, I have developed a more intimate relationship with the things themselves. This has led me to contemplate their purpose and my responsibility to the choices I have made to be their guardian, whether that's the blackberry I hold for a few moments before eating it or the favourite knitted pair of socks that I keep darning over holes that appear through wear and tear to last a lifetime.

The more I have researched materials and their environmental impact, the more obvious it becomes why big corporate brands choose materials that inherently break or wear down easily. It traps the customer in needing to buy new items. To mend and tend to your items is an act of resistance to these corporations which are hugely responsible for the devastation that consumerism has had on our planet. As always, it can't only be down to us as individuals to take action, and not everything can be mended. It simultaneously takes policy, such as the Right to Repair Act, which is intended to allow customers to repair and modify their electronics rather than only using the brand's services. This act was first implemented in 2012 in the motor vehicle industry, and while it may seem like an obvious consumer right, companies such as Tesla and Apple have found loopholes in the law, allowing them to bypass this and other such acts. This freedom to mend and tend to our possessions and goods should be available to us.

When items are made of quality materials and are taken care of, they can last a lifetime and be passed down through generations, donated or resold. Storing things correctly can extend their usefulness as well; properly stored fruit and vegetables might last ten days rather than two. Extending the use of one item of clothing for just nine months reduces its carbon, waste and water footprint by around 30 per cent.

If you can make something last longer, you are able to significantly lower the impact of that product on the planet and extend your enjoyment of it.

KITCHEN SOLUTIONS

There are some items in my kitchen that I consciously try very hard to make last but find too hard to resist, such as baked goods or fresh fruit. The majority of items I actively work to keep fresh for longer. I hate to waste food, which was something drilled into me long before I understood the environmental impact of food waste. My parents would always say, 'Waste not, want not', which I never quite understood as a child. The way I now understand that phrase is, 'Don't buy it if you're not going to consume it and let it go to waste'.

I grew up in the heart of London. On my parents' journey home from work, they would walk past about four supermarkets, so they would usually buy small amounts of groceries more often. This was beneficial to our lifestyle, as they didn't need to plan too far ahead, and if plans changed, it was likely that not much would go to waste in the fridge. I now live in a more suburban neighbourhood, and I certainly don't pass many supermarkets while walking, so when I go shopping, I buy more and go less often. This has meant I've had to develop more skills to preserve my produce. Here are some tips to reduce food waste:

- ⊙ First, don't buy more than you need.

- ⊙ What you do buy, store well to keep it fresh.

- ⊙ Make sure you use all edible parts of your vegetables and fruit.

- ⊙ And lastly, compost! (See page 222 for a guide to composting.)

NO MORE WILTING HERBS

I really enjoy cooking with fresh herbs, but as they are typically sold in huge bunches, I find it challenging to use them all before they go bad. A chef once taught me this super easy way to keep herbs and leafy greens fresh.

Wet a clean tea towel and squeeze out the excess water (you want it damp rather than dripping wet). Then lay it out on the worktop, place your herbs on top and roll them up like a little parcel. You want to make sure there are no gaps and the towel is fully covering the herbs. Store in the fridge and use the herbs as you need them. Each time you rewrap the bundle, check that the cloth is still damp and, if not, run it under the tap once more.

This kitchen storage tip has saved countless bunches of herbs and greens from going to waste. If you want something a little more purpose-designed, there are also highly effective storing bags made from absorbent fabric such as terry towelling for storing delicate greens.

VEGETABLE VASES

Some vegetables and fruit are best stored submerged in water. Similar to a bouquet of flowers, they like their water refreshed every few days. Depending on the shape and quantity of what I am storing, I use a mixture of bowls, glass jars and reusable containers to hold the water and vegetables. For tall veggies such as celery, submerging even half of the stem is more than enough to keep them fresh. The following vegetables all like to be stored this way:

- Carrots (completely submerge)
- Radishes (completely submerge)
- Asparagus (partially submerge)
- Spring onions (partially submerge)
- Celery (partially submerge)

FREEZER STORAGE

I used to associate the freezer with shop-bought precooked food and ice cream, but later in life I discovered it was an incredible tool for saving ingredients and meal prepping for those days when you have nothing fresh to eat. Here are some items I love to prepare ahead and freeze:

- Steamed vegetables in portion sizes (a large ice cube tray works great)
- My go-to veggie burgers (see page 133)
- Sliced loaves of my sourdough bread
- Overripe fruit and avocados (these are great for smoothies)

I like to do a freezer 'spring clean' every couple of months to make sure I am not forgetting about items in the back. How full my freezer is usually fluctuates, so I adjust the freezer temperature accordingly to avoid wasting energy.

ARCHITECTURE OF A FRUIT BOWL

There is nothing more disappointing than reaching towards your fruit bowl, picking up a piece of fruit you were looking forward to eating, and finding it has begun to rot. This might be because you have truly left it too long to eat it or, more likely, it's because you are storing some of your fruit together that would be best stored apart. I was always curious as to why certain fruit ripened differently and, if it was touching another fruit, how that would also affect the process. That curiosity led me to learn that fruits have different levels of ethylene, which is best described as a fruit-ripening hormone. If you design the layering of your fruit bowl or, even better, have separate fruit bowls for the different levels of ethylene, your fruit will ripen more evenly.

FRUITS THAT RELEASE HIGH LEVELS OF ETHYLENE		FRUITS THAT RELEASE LOW LEVELS OF ETHYLENE	
Apples	Mangos	Avocados	Oranges
Pears	Plums	Lemons	Grapes
Bananas	Nectarines		

Ethylene in general ripens food faster, so if you prefer to store some of these fruits in the fridge, be careful not to have them next to your vegetables. On the flip side, these high-ethylene-releasing fruits can be used as a trick to fast track the ripening of other produce, if desired.

When considering where to place your fruit bowl in the kitchen, keep in mind the locations of your windows. While fruit does look particularly photogenic in an area that receives a lot of sunlight, your fruit will ripen more evenly if the bowl is placed in a shadier spot. I am a huge believer in the Mediterranean way when it comes to fruit and some vegetables, and I try to store everything I possibly can outside of the fridge. The room temperature and air flow allow them to ripen naturally, and with ripening come a development and complexity of flavour.

SHADED VEGETABLES

There are a few vegetables, such as my favourite, the potato, that are best stored in a cool, dark place, such as a cupboard that doesn't receive a lot of direct sunlight and isn't by your oven. The potato has fed civilizations through famine because, when stored correctly, potatoes can stay edible through the long winter months. Other vegetables that like being stored like this include sweet potatoes, garlic, onions and all varieties of winter squash such as butternut squash and pumpkin. You also want to store these vegetables in containers with airflow, such as a basket, cardboard box or a mesh cloth bag. This method of storage can extend the life of these types of vegetables from two to three weeks to two to three months.

PANTRY STORAGE

The glorious and glamorous-sounding pantry. When I refer to my pantry, I want to be clear this is not a palatial walk-in room dedicated to dried goods, although I do daydream about such a place more than a walk-in wardrobe. I like to imagine all my beloved grains, beans, nuts, seeds, spices and flours beautifully organized. For now, I have a somewhat more chaotic cupboard full of a broad assortment of upcycled jars with all different shapes, sizes and coloured lids. I like to put my dry foods in jars because I can clearly see what they are, and it's easier to access them and keep them airtight.

It's always good to do research when considering the shelf life of items. Items that you might assume have similar shelf lives might not. For instance, white rice can last almost indefinitely when stored in an airtight container whereas brown rice is good for only three to six months. If you have decanted the food from a package into a jar, write the best-by date from the package on the jar with a wax pencil, or cut that section of the label out and put it into the jar.

A frustrating pest in the kitchen is the Indianmeal moth, more commonly known as the pantry moth or weevil moth. These insects love to feed on flours and grains, which is terrible news for our pantry goods. You can deter these insects by soaking a cotton-wool ball in peppermint oil and putting it in a small dish inside your pantry cupboards. The Insect-Repellent Wardrobe Ball on page 187 can be used in the pantry, too.

UPCYCLING JARS

There is no need to go out and buy mason jars or cute matching containers when you can save and upcycle jars and containers from shop-bought items such as jams, sauces and condiments. The only jars that I have bought are extremely large sizes that products are usually not sold in. Most jars have these sticky labels that are challenging to remove. I will admit I am often lazy and leave them on, but when I do carve out the time for a little jar cleaning, it's similar to the feeling when you wash your windows and suddenly the view is superior – the view here being the delicious food you are storing inside the jar.

Here's how to clean your jars:

① First make sure you get everything out. I find a silicone spatula works really well to clean out my jars. Then wash the jar with soap and a brush.

② Fill a large bowl with hot water.

③ Fill the jar three-quarters full of hot water and screw the lid back on.

④ Place the jar inside the bowl and leave to sit for about 30 minutes or longer. You will see the label start to come off as well as the adhesives. It may need a little help with a scrubbing brush and some washing-up liquid to get it extra clean.

⑤ Drain the water and dry the jar before using as a storage container.

WARDROBE SOLUTIONS

Fashion is one of the most polluting industries on the planet, so making an effort to not only buy responsibly but also to take care of the items already in our dressers and closets can have a big impact. With a few tricks and some basic sewing skills, you can easily preserve, repair and even breathe new life into your favourite garments.

CLOTHING STORAGE

You may have items of clothing that you don't wear in some seasons, such as a winter coat. By storing these in opaque clothing bags when you are not wearing them, you can prevent sun exposure they may experience in an open wardrobe. Clothing bags will also help deter fabric-eating bugs such as moths. Before storing items, it's good to wash them to rid them of any dust or bacteria. I experience very short winters in California, so I like to store away the winter coats during the long summer, along with my beloved beanies and scarves.

MAKES 1 LARGE REPELLENT BALL

Lightweight cloth fabric, at least 30 x 30cm (12 x 12in)

30cm (12in) diameter (or larger) circular tray or plate

Scissors

Dried herbs, cedar woodblocks or shavings, or a cotton-wool ball soaked with essential oils

Ribbon or string

tip • It is great to clean out your wardrobe thoroughly as the seasons change. You can wash and refill your insect-repellent bags and consider different mixes of oils and leaves to mark the new season and continue to keep your clothes free from invaders.

Insect-Repellent Wardrobe Ball

It is so upsetting to reach for a favourite jumper and find the moths have been there first. Using this homemade repellent, you can help keep your wardrobe and drawers free of moths and other clothes-damaging critters. In bygone days, people stitched little bags of these leaves and wood shavings into their clothing to both smell sweeter and repel bugs. To protect your clothes, make some similar bags from scrap fabrics or old T-shirts and saved ribbons and string, no sewing required. These also make great gifts if you choose pretty fabric.

There are several aromas that bugs dislike, including eucalyptus, cedar, peppermint, lavender, cloves and rosemary. You can make this with dried herbs from the supermarket or farmers' market, or you can grow and dry them yourself. Rosemary and mint would be a good place to start as these are very easy herbs to take care of.

Place the plate on your fabric and trace around it with a pen or pencil. Cut out the circle and fill the centre of the fabric with some dried lavender, cedar woodblocks or shavings, eucalyptus leaves, dried mint or a mix of these (you want about 60g/2oz total). Bring the edges of the fabric together and tie with ribbon or string, then tie it to the rail in your wardrobe. Alternatively, you can place a cotton-wool ball soaked with a few drops of essential oils in the bag. Smaller sachets can be made for your drawers. •

REPLACING A BUTTON

This is one of the many things they should teach you at school! How great would it be if simple home economics skills were in school curriculums – such as sewing on a button, hemming a pair of jeans, or personal financing skills such as budgeting your food shopping.

It is frustrating when you have a well-loved shirt missing a button sitting in your wardrobe not being worn because you keep forgetting or don't know how to replace it. You can sew on the extra button your garment likely came with, or you can switch out all or some of the buttons to give it a fresh new look. The buttons on a garment do not need to match; try elevating a plain garment with a wild mix of buttons. Beautiful vintage buttons can often be found at flea markets and secondhand shops. Start making a collection for your button box; each one tells a story, a narrative you can weave into the garment to which you sew it.

Here's how to replace a button:

① Mark the spot where the button is to be sewn. Thread your needle and tie the two ends together. Pass the needle through the spot from the back to the front through a hole in the button, then back again through another hole.

② Then pass the needle into the third hole from the back to the front.

③ Pass the needle through the fourth hole from the front to the back.

Repeat these processes five to six times.

④ Pass the needle through the button but not through the fabric, and wrap the thread around the stitches two to three times.

⑤ Pass the needle through to the back of the fabric and tie off by making a few small stitches.

⑥ With a four-hole button, you can sew through the holes in pairs, or I prefer the detail of passing the thread to the diagonal hole to form a cross.

If you can't find the time, take your item to your local dry cleaner and utilize their expertise – just don't forget to request no plastic sleeve.

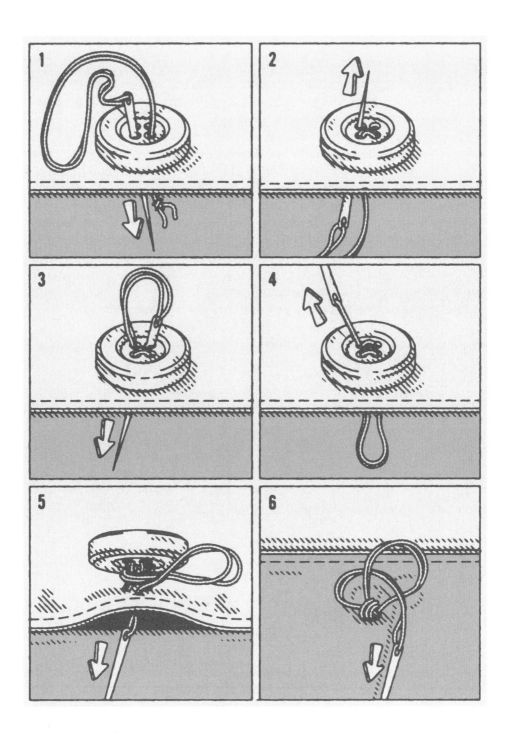

DARNING, FROM ART FORM TO NECESSITY

My mum is a brilliant sewer, and while I will never have her skill, I do share her opinion on the importance of mending items and her imagination on how to make things new. Repairing by darning or sewing is an effective way of prolonging the life of a favourite garment that has holes or signs of wear and tear.

Recently, a lot of textile artists have included darning and repairing in their practice as a statement about waste and consumption. This takes reference from the Japanese art of *kintsugi*, meaning golden joinery, which, through the technique of applying a seam of lacquer and precious metal, adds a new dimension and visual interest to a cracked or broken vessel. I love when darning deliberately draws attention to the repair through the use of contrasting colours or embroidery stitches. Jeans are a great place to try out this technique – when you stitch on a patch to repair a tear, the stitching can either blend subtly with the denim or make a louder statement. Holes in socks can be easily darned by using contrasting yarn as a statement or invisibly with the same colour.

Not only is mending a practical skill that will save you money and lower your carbon footprint, I also find it to be a wonderfully meditative process – the regular back and forth of the needle with the end satisfaction of a hole filled, a rip repaired or a button replaced.

LAUNDRY SOLUTIONS

A lot can happen in the laundry process that can affect our waterways, our soil and eventually us. While most cities have water treatment plants that treat household wastewater before it returns to the ocean, there is only so much they can truly filter. We have the power to make choices at home to safeguard the health of our waterways and ourselves, as it all flows back to us.

MICROPLASTICS

The most shocking thing that happens when we wash our clothes is the amount of microplastics that can shed during the process. Microplastics are teeny, tiny pieces of plastic that come off clothing that contains synthetic fabrics, such as nylon. The microplastics drain through our washing machines and into our waterways. What can you do to reduce this pollution?

Use products that capture microplastics: there are lots of brands creating clever tools to capture the microplastics shed during the washing machine process and more safely dispose of them in your household rubbish. They include the Guppyfriend washing bag, the Cora Ball and the Filtrol microplastics filter.

Wear natural fibres: a more long-term approach would be to reduce the amount of microplastics that enter our waterways by buying clothes made from natural fibres, such as cotton, hemp, wool and linen. The fibres these textiles shed will biodegrade and not pose a threat to our oceans.

LAUNDRY LOAD TIPS

Other things to consider when it comes to your laundry:

- Wash your clothes on a cold or cool cycle. Ninety per cent of the energy required to run a load of laundry is used for heating water, so switching to cold wash can make a huge difference in your energy use.

- Wash larger loads so you are washing less often.

- Ask yourself, Does this really need washing yet? Or could you get another wear out of it?

- Seek out laundry detergent with plastic-free packaging. There are companies such as Smol that ship plastic-free pods in cardboard boxes, or you can find other laundry products at refill shops.

- If plastic-free is hard to find, look for larger bottles of detergent so that the product-to-packaging ratio is higher, resulting in less package use in the long term. Or choose concentrated formulas that can be diluted.

- Could you dry your clothes using air and sun either on a rack inside or a clothesline outside rather than using a dryer?

- Ditch the dryer sheets. They end up in landfill, where the nonbiodegradable chemicals found in them leach into the soil.

- Use wool dryer balls. Dryer balls help absorb water from the clothes and speed up drying, reducing energy consumption by 10 to 25 per cent. You can also put a few drops of essential oils on the balls to make your clothes smell nice.

STAIN REMOVER

I am a pretty messy person when it comes to stains. I either have a hole in my chin or a bad sense of spatial awareness, as I am constantly needing to remove stains. I also love to wear white trousers and have a dog, so I am asking for it.

The minute you stain your clothes, apply salt to soak it up. Then, once you are not wearing the item, apply lemon juice directly to the stain. If it's a tougher stain, mix lemon juice with bicarbonate of soda and apply to the stain, then rinse with cold water and wash as normal. I have also found some great stain-removing soaps that come free of packaging or in recyclable cardboard.

BRIGHTEN-YOUR-WHITES SOAK

Over time, white clothing and linens can lose their brightness. This soak is great for white T-shirts, bedsheets and napkins.

Fill a large pan with water and add three lemons, sliced. Bring to the boil, then turn off the heat and add your clothing or linens. If they're delicates, leave the water to cool a few minutes before putting in the fabric. Leave the items to soak for an hour and then wash as normal. After washing, dry them in the sun – this saves energy by not using the dryer, and the sunlight naturally helps to brighten the fabric.

ENERGY-SAVING TIPS

The average home or apartment uses a lot of energy! In this day and age, we charge countless devices, heat large spaces and run multiple appliances, sometimes all at the same time. It can feel almost pointless to try to curb energy use in this era of mass consumption, but a great place to start is with these simple, very manageable steps that really do make a difference.

ELECTRICITY

- Switch to energy-saving bulbs. LED light bulbs are 75 to 80 per cent more energy efficient than incandescent bulbs. Look for bulbs that have EU Energy labels on them and assess their energy efficiency.

- Unplug anything not being used. Even when appliances aren't in use, they can continue to draw power.

- Only run appliances such as dishwashers when they are completely full. I also experimented with my dishwasher and mainly run it on a light cycle when just washing plates, cups and utensils, as it reduces the cycle time and still does the job, but again, each machine is different.

HEATING

- Wear weather-appropriate clothes even when you are indoors. If you dress for summer inside during the winter, you will need to keep your house very warm to be comfortable. Layer up before raising your thermostat.

- Install a smart thermostat if you are able to, as they can more accurately sense the temperature in the house and know when to turn the heat up or down.

- A lot of heat can be lost through drafts in the home from gaps under doors or windows, for example. Walk through your home and see where heat, and therefore energy, could be escaping or cold air coming in. You can reduce wasted energy by using draught excluders; if the issue is larger, the door or window might need mending.

- Windows let in both a lot of cold and a lot of heat. In the winter months, close your curtains or blinds straight after the sun goes down to trap in the heat from the day. In the summer, do the opposite, keeping curtains and blinds down during the day in rooms or areas you are not using, and partially shut in the rooms you are using. Open them when it gets cooler at night, to let the cool air in and let built-up heat escape.

WATER-SAVING TIPS

⊘ Every minute you run a shower, you are using approximately 7.6 litres (2 gallons) of water. If your shower takes a while to heat up, capture the unused water in a bucket. This can be used to water plants, clean the floor or hand-wash clothes.

⊘ When washing your hands, wet them, turn the water off while you are lathering and then turn it back on to rinse.

⊘ The same goes with teeth brushing. Use a little water to wet your brush, then turn off the water while brushing. This can save up to 11.4–15.2 litres (3–4 gallons) of water per person per day. Annually that's equivalent to the water used for one hundred four-minute showers.

⊘ When hand-washing dishes, rather than letting the water constantly flow as you wash, do a quick rinse of any heavily dirty dishes, then fill up the sink with warm soapy water, wash everything at once and rinse the soap off all together afterwards.

⊘ Think twice before you flush. Toilets account for nearly 30 per cent of an average home's indoor water consumption. That's a lot of water! By only flushing the toilet every third or fourth time you pee, you can save up to 57 litres (15 gallons) of water per day.

o o o

All in all, there are many tips and tricks that we can implement into our daily lives to help extend the life of our things and lower the costs of keeping a home. Just like our bodies need tending and nourishing, our possessions need the same long-term perspective and level of care. We are noticing as a society that we have been profligate consumers of energy, and as we radically rethink systems of consumption and extraction on a large scale, it's important to remember all the seemingly small changes we can make within the home that contribute to this global movement. Resist the urge to immediately throw something away when it has the smallest imperfection, give your things the chance to be fixed and evolve, and you, too, will grow.

go make

HOMEMADE PRODUCTS TO
NOURISH YOURSELF AND YOUR HOME

I love to make things. I have always been someone who finds great calm in doing things that I can feel and touch. I get this from my dad, Gary, who is a goldsmith, and his dad, who was a carpenter. My dad loves to make and create things, and everything is an art project. We have enjoyed making many homemade projects together throughout my life. From carving faces in wine bottle corks to make toys, to making metal spoons, to staying connected and creative by drawing portraits of each other over video calls.

The world can often feel so abstract and out of reach; I find reconnecting with materials and tangible processes is a great exercise to promote well-being and peace. As I have experimented with different solutions to be less wasteful and more resourceful, I found that by making certain items from scratch using simple, natural ingredients, I experience the soothing benefits of working with my hands and end up with a more affordable, often superior product. By applying my imagination, a resourceful attitude and some patience, I can make the products I need.

Just as mending and tending to your possessions can be an act of resistance and reflection, I find making things yourself can be, too. Even the small act of making my own toothpaste or growing a plant can help me feel empowered and more self-sufficient. The items that we use every day in our home environment are incredibly intimate and personal, and I can't think of a better way to celebrate this relationship than by making them myself.

BATHROOM SOLUTIONS

Let's begin in the bathroom. I used to think that making beauty and hygiene products required laboratory-like facilities and was impossible to do at home. Perhaps I was led to believe this due to the ultra-clean and clinical marketing strategy of so many brands. When I started shifting to more simple and natural products, I found that these products were made with ingredients I recognized rather than long chemical names that I couldn't pronounce. From there it was a small step to the realization that I could make them myself. While I haven't gone as far as distilling essential oils yet, I have been surprised at how many simple items you can make for your bathroom needs, from the essentials to more indulgent products.

Earth Paste

3 tablespoons coconut
oil

3 tablespoons
bicarbonate of soda

2 tablespoons bentonite
clay (optional)

1 tablespoon
diatomaceous earth
(optional)

30 drops peppermint
essential oil

tip • In the colder months,
you might find your toothpaste
becoming rather solid, as
coconut oil hardens at cooler
temperatures. I like to use
the end of my toothbrush
to scoop it out or to store
a little spoon next to it for
scooping. Once you decide
you like the ingredients and
find your perfect ratio, you
can buy bigger bulk-size packs
of the ingredients for a lower
packaging-to-product ratio in
the long run.

Let's start with an essential item you use at least twice a
day: toothpaste. This was the first bathroom DIY item I
attempted to make at home, and the first two batches did not
taste good! But the experimenting was all part of the fun and
discovery. I started by making a very small amount at first so
I didn't waste the ingredients. Where possible, try to seek out
the ingredients in recyclable or reusable packaging materials.

Food-grade diatomaceous earth and bentonite clay, which
can be found at garden centres or DIY shops and large
supermarkets, respectively, are the ingredients I added as I
played with this recipe. These two ingredients are optional,
but I found they add a soft abrasive quality to the toothpaste
and create a thicker texture, which I preferred. Toothpaste
can be very personal, so if this ratio isn't to your liking, play
around with it.

Heat the coconut oil in a saucepan over a low heat just
until it melts; you don't want it to boil. Pour it into a small
glass container, such as an old jam or condiment jar (this is
also the jar you will store it in). Leave the oil to cool for 30
seconds, then add the bicarbonate of soda and, if using, the
bentonite clay and diatomaceous earth. Mix thoroughly, then
add the peppermint essential oil. Store at room temperature
for up to 6 months. •

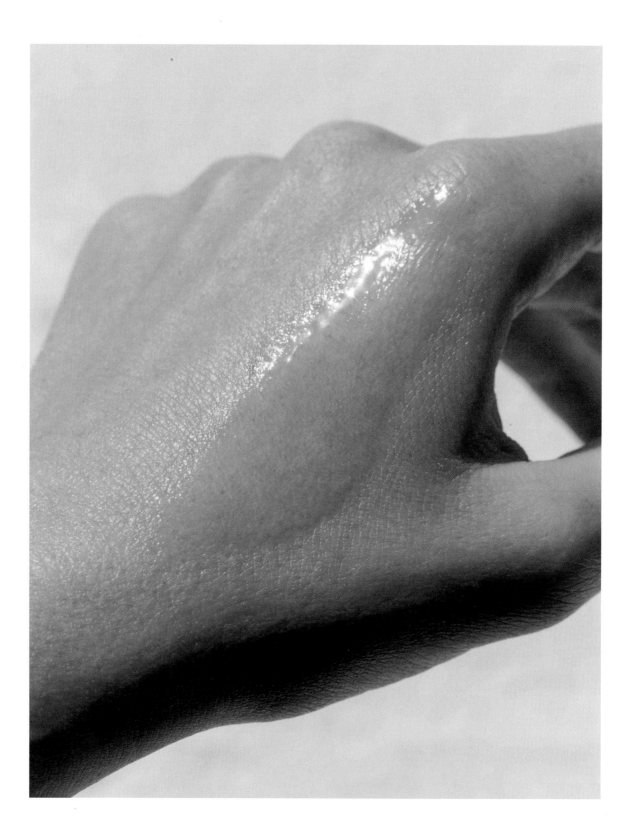

Allover Balm

**3 tablespoons grated
cocoa butter**

1 tablespoon olive oil

**A few drops of essential
oil**

tip • I once made this with peppermint oil because it was all I had at the time, and whoa, was it fiery on my lips. So I wouldn't suggest that as your essential oil unless you like that minty bite! Now that you have bought the cocoa butter, one thing you must try is using it to make your own chocolate. Just melt it, then stir in cocoa powder, maple syrup or honey and a pinch of salt. I store mine in ice cube trays in the freezer.

Body balm, lip balm, hand balm, it's all the same. A simple way to lower the consumption of products – often marketed in a very seductive way by the beauty industry – is to reduce the number and variety of products you use. This allover balm replaced at least four different items on my top shelf, and I find I only need about half the amount of those products to get the job done! My skin soaks it up, and I have found it particularly helpful to keep my skin hydrated when travelling and in the winter months. I like to make this in a big batch, store the bulk of it in a jar and then have a small transportable container to keep with me at all times.

Fill a saucepan with 5cm (2in) of water and place a heatproof bowl over the top to create a double boiler. Place the cocoa butter in the bowl and heat the water over a medium to high heat until the cocoa butter has completely melted. Carefully take the bowl off and place it on your worktop. Stir in the olive oil and essential oil, then pour the mixture into a clean tin (I usually use a reused lip balm container) and leave it to cool overnight. •

Bath Salts

MAKES ENOUGH TO
FILL A 1 LITRE (34FL OZ)
MASON JAR

600g (1lb 4½oz) Epsom
salts

200g (7oz) coarse sea
salt

40g (1½oz) bicarbonate
of soda

20–30 drops of essential
oils (optional)

A handful of dried
flowers or herbs
(optional)

tip • If your bath salts get
too damp and start clumping
together, poke holes into the
lid of the container (if it's a
metal or plastic lid) so the
salts can breathe. I have found
using a container with a cork
lid works really well to keep
bath salts from clumping.

A bath is like a hug that you get to enjoy all to yourself.
These salts make a great gift! This simple three-ingredient
base can be customized with whichever essential oils and
dried herbs you like. My favourite scents that I will almost
always use are eucalyptus and lavender, but mint or rose also
work great here.

Combine the Epsom salts, sea salt and bicarbonate of soda
in a medium bowl and stir together with your hands or a
spoon. Mix in the essential oils and dried flowers or herbs,
if using, then decant into a mason jar or clean upcycled
container. I usually use about two handfuls per bath. •

Calming Green Tea Face Cloths

1 tablespoon loose-leaf green tea (or 1 tea bag)

2 cotton washcloths

tip • I usually make five to seven wipes so I have enough for the week and scale the recipe accordingly. There is no need to buy new washcloths to do this. You can use any thick piece of cotton fabric, or cut an old towel into squares.

A couple of years ago, I stopped using disposable face wipes and switched to washable reusable cotton face rounds for my eyes and cotton washcloths for my face. Disposable face wipes are designed to easily and quickly remove makeup and dirt, but this quick-and-easy product lasts a lifetime on our planet. It can take at least one hundred years for a face wipe or baby wipe to decompose. I find they can also be pretty abrasive on my skin. Before making the swap to reusables, I liked a brand of green tea face wipes, so I decided to make my own.

Make a pot of green tea and leave it to steep for 8 minutes. Pour the tea into a bowl, add the washcloths and leave them to soak for 10 minutes. With clean hands, gently squeeze the washcloths, getting most of the liquid out but still keeping them relatively damp. Fold the cloths and place in an airtight container. I like to store mine in the fridge to keep them extra cool and calming when I use them on my face. •

Coconut Coffee Exfoliator

**A palmful of cooled
coffee grounds (about
40g/1½oz)**

**1½ teaspoons coconut,
olive or vegetable oil**

tip • Try to keep your coffee
grounds from going down the
drain and instead compost
them. I have found a mesh
filter over the plughole works
well for this.

If you're like me, you go through a lot of coffee! Hopefully
by now you have started composting your coffee grounds or
maybe you're thinking about it. One morning I was emptying
my fabric reusable filter of coffee into my compost bucket,
and I thought surely there is something more I could do with
these grounds. They smell *so* good, they are rich in nutrients
and they have naturally occurring oils. It was the oil in the
coffee that made me think they could be moisturizing, and
sure enough, when I mixed the grounds with some coconut
oil, the result was this perfect moisturizing and exfoliating
product. I also love to scale this recipe, multiplying it by four
to make enough for an allover body scrub. I advise you to do
this standing on a towel as it can get messy.

Mix together the coffee grounds and coconut oil in a small
bowl. Rub the two together in your hands over the sink –
do that for a minute or so, really getting into the areas that
feel tight and relieving pressure. Once you have shown your
hands a good amount of love, wash them and then dry. I
hope they feel silky soft! •

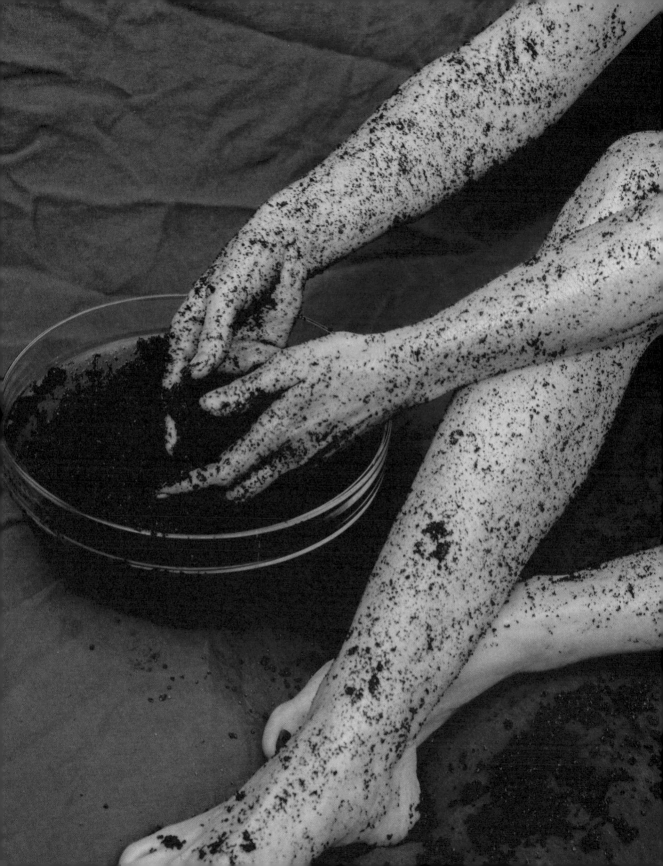

KITCHEN SOLUTIONS

I'll admit, the idea of making washing-up liquid and other kitchen products didn't have quite the same initial appeal as making over my beauty cabinet. Reworking this section of the home felt less indulgent and maybe a little less fun, but once I got into the swing of things, I found it extremely satisfying. Products such as cleaning spray and cloths for wiping down surfaces get a lot of use in our kitchens and often come heavily packaged, so switching to upcycled homemade versions can make a big difference.

Washing-up Liquid

MAKES ENOUGH TO FILL AN OLD WASHING-UP BOTTLE (ABOUT 540G/19OZ)

3 tablespoons grated Castile soap (I use Dr Bronner's mild baby soap)

½ teaspoon washing soda (sometimes called soda crystals; found at DIY shops or online)

40 drops essential oils (I love to do a citrus-and-herb blend such as lemon and thyme)

tip • Washing soda is also great for adding to loads of laundry to remove stains and brighten your clothes, to unblock drains or even remove grease from surfaces, pans and grill grates.

Out of all the household tasks, I have always enjoyed washing the dishes. I remember I would jump up at my Nanna Beryl's house to do the dishes because I loved that she had a double sink! She would always show me that you can save water if you filled one sink up with warm bubbly water and used the second sink to rinse the dishes and leave to dry. On reflection, I love that she said that, because it made me be more conscious of water usage. I remember I used to stand up on a stool to reach the sink, and I loved that my little arms could be submerged in water. A bubble bath for my hands! Not sure how well I actually cleaned the dishes. . .

In a saucepan, heat 960ml (32fl oz) water, the Castile soap and the washing soda over a medium heat. Mix with a wooden spoon and slowly bring to the boil. Once everything has dissolved, remove from the heat and leave to cool to room temperature. Add your essential oils, decant into an upcycled washing-up liquid bottle and leave to rest overnight before using. Don't be alarmed if this gets thick and cloudy and even clumps up a bit as it sits! Just give the bottle a good shake and keep washing. •

Multifunctional Citrus Cleaning Spray

MAKES 1 LITRE (34FL OZ)

480ml (16fl oz) white vinegar

Peels from about 6 lemons and/or oranges

A few sprigs of thyme (optional)

Spray bottle

Essential oil of choice (earthy scents such as thyme, rosemary or mint work best)

tip • Rather than buy machine-washable cleaning cloths (and please say goodbye to disposable paper towels!), cut up old towels or T-shirts to use for cleaning.

I call this 'multifunctional' because cleaning supply companies have convinced us we need a different spray for every room. Why do I need one cleaning spray for the kitchen and another one for the bathroom? You won't fool me into buying the same thing packaged in two different bottles anymore. I love the citrusy smell of this spray, and it's such a clever way to give a second life to used up orange and lemon peel. While I do consider this an all-purpose cleaner, this spray should not be used on wood or granite.

Add the vinegar to a 1 litre (34fl oz) mason jar and tuck it into your pantry. When you eat an orange or squeeze a lemon for something, toss the peel in the jar and replace the lid. Once your jar has at least six peels, throw in a couple of sprigs of thyme (this is optional, but I always add it as thyme is said to have antibacterial qualities), reseal the jar and store in a cupboard for 2–3 weeks.

Once the vinegar is infused, strain out the peel and herbs. This is now your cleaning concentrate. Fill an upcycled cleaning spray bottle or a new glass bottle one third with this concentrate and then fill to the top with water to dilute. Finally add 5–10 drops of essential oil. •

80g (2¾oz) bicarbonate
of soda or washing soda

60ml (2fl oz) white
vinegar

tip • You can also mix this
solution in a bowl or jar
and apply it to hard-to-clean
surfaces such as bathroom
tiles. The best tool for getting
into all the little crevices of a
bathroom is an old toothbrush.
The solution can even be used
to clean your washing machine
and kettle.

Sink Freshener

I like to think of this as a less intense, more ecologically
friendly version of the chemical-laden clog-removing
products lining DIY shop shelves. By simply mixing together
two pantry staples, you can easily brighten your sink basin
and clean out your drain – no scary chemicals required.

First, pour hot water into your sink and drain. Shake over
the bicarbonate of soda, applying it to the areas you are
working to treat – that could be the sink and drain or just
the drain. Then pour in your white vinegar. You will see a
bubbling effect like a school science project. Leave it to sit
and do its thing for as long as possible, then rinse with hot
water again. •

FABRIC-BASED SOLUTIONS

Next up, let's transition into the somewhat softer, more gentle space of fabrics. I called on my mum for some design help with these next three items. For my thirtieth birthday, my parents gifted me a sewing machine and, ever since, my mum and I have been having sewing classes over video call. Living so far away from each other, it has been an enriching and joyful way to share time and learn skills from her.

If you don't already have a small sewing kit, I highly recommend putting one together with the basics. You can find these materials in a craft shop or online. I keep my kit in a sweet little cylinder box that a gift once came in. Here's what you'll need:

① Assorted thread (your colour range will develop over time)

② Needles

③ Straight pins

④ Measuring tape

⑤ Small scissors (12.5cm/5in) to snip threads

⑥ Fabric shears (23cm/9in or longer)

⑦ Buttons (your collection will expand over time – I always save the spare ones that come attached to clothing)

Reusable Coffee Filter

Paper coffee filter

30 x 15cm (12 x 6in) piece of medium-weight cotton fabric (I repurposed one of the many tote bags I had; just make sure the fabric is clean)

Pencil

Scissors

Needle or sewing machine

Thread

tip • When you clean your filter, avoid using washing-up liquid, as you will find your coffee will soon start tasting of it. I like to do a thorough clean of my filters once a month by heating them up in boiling water and letting them soak for an hour with the heat off.

In the scheme of things, a reusable coffee filter may seem like a small contribution to reducing waste, as paper and bamboo ones are readily available and compostable. However, 2.25 billion cups of coffee are consumed globally each day. If one-third of those cups are made with paper filters in 2 litre (68fl oz) batches, 750 million paper filters are being made and discarded each and every day – 275 billion per year. That is a lot of trees!

A reusable filter is a great way to make your caffeine routine less disposable and save money. I have been using a cotton filter for a couple of years, and it makes great, clean coffee without sediment. Here is how to make one yourself, sewing either by hand or by sewing machine.

① Fold the fabric in half.

② Using a paper coffee filter as your template, place one of the sides of the paper filter template against the fold. Draw around the template.

③ Cut out the fabric along the line.

④ Unfold the fabric and sew along the top curved edge to prevent fraying. Fold the filter back in half and stitch up the open side and the bottom.

I hope that the disposable filter you use here as your template is your last! It's not a bad idea to make two filters while you have your materials out. •

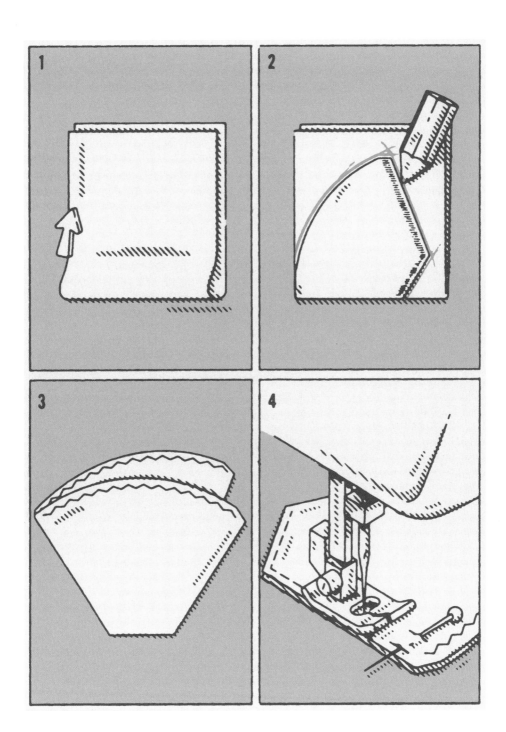

Reusable Tea Bag

7.5 x 30cm (3 x 12in) piece of medium-weight cotton fabric (make sure it's clean if you're repurposing fabric from something else)

Needle or sewing machine

Thread

tip • You could even consider planting your own herbs for infusions. Mint, for example, is not only delicious for tea but is also a very easy and fast-growing herb to grow yourself.

When I discovered that a lot of tea brands now use plastic in the material of the bag, the tea-loving English in me was devastated. Would this mean I have to give up tea? Firstly, not all tea bags have plastic in them. I would suggest you do your research, check with the brand you buy, and call them out if they do. They want to keep their customers! But this discovery actually led me on a beautiful journey through the world of loose-leaf tea and the joy of creating my own blends. I also learned that in the same spirit of the reusable coffee filter, you can easily make little bags for loose tea or herbal infusions. Happy brewing!

① To prevent fraying, put a line of stitching on the two short 7.5cm (3in) sides of the fabric.

② Then fold in half by bringing the two short edges together.

③ Stitch each of the long sides together, leaving the top open.

④ Fill with loose-leaf tea and enjoy. •

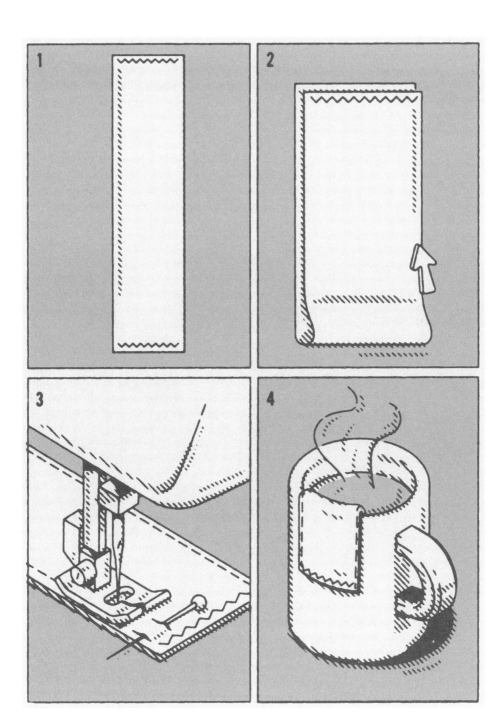

Natural Fabric Dyes

Cotton or linen fabric,
such as napkins, T-shirt
and socks

Large cooking pan

12 avocado stones (you
can also use the skins)

Spoon or tongs

tip • There are many variables
in dyeing. Every avocado is
different, as is the colour its pit
produces. Soaking time also
dictates the colour, vibrancy
and shade. How the colour
takes also varies massively
with the type of fabric. I just
played with cotton, but it's fun
to experiment with different
types of fabric once you have
the hang of it.

A great way to brighten your clothes is by dyeing them –
this could be bedsheets you've had for a while that need
revitalizing, a pair of white socks you want to jazz up or
some napkins you want to dye and gift to a friend for their
birthday. There are many items from nature that can be
used to dye fabric instead of using synthetic dyes that can be
harmful to our water systems and our skin. Many of these
natural dyes also don't require a dye fixative, a chemical
compound needed to set the dye so it doesn't wash out.

The first natural dye I practised with was from the stones
of avocados, and I was quickly mesmerized by the soft
dirty pink hue it imparts. To me, this soft yet earthy colour
epitomizes the tone of *go gently*. Other vegetables can be
used for various colours, but avocado is still my favourite.
Let's try dyeing with them first, and then you can explore
as many shades as you like. To give you an idea for material
scale, I am going to dye a set of six napkins, a T-shirt and a
pair of socks. Once you have all the materials out, I find it's
best to dye a few different items in one go. I like a medium
to light pink tone; if you'd like something darker, use more
avocado stones.

First, wash and prep the items you are dyeing. Wash them
on a normal cycle, then soak overnight in warm to hot water
with a small amount of gentle laundry detergent. The next
morning, give them a good rinse. This prep helps ensure the
dye takes to the fabric evenly.

Fill your pan with water; you want to use a big enough
pan that your items can move around freely. Wash your
stones, making sure there is no residual avocado fruit on
them. I like to clean the stones and store them in the freezer
until I have enough to dye with. Now add the stones to the
pan and bring the water to the boil; a gentle boil is enough
rather than anything too vigorous. Turn the heat down and
simmer for 30–60 minutes, or until the avocado stones begin
to turn the water a beautiful deep murky pink.

Now add your fabric items and keep at a low simmer for 15 minutes, stirring with tongs or a spoon a couple of times. Then turn the heat off. How long you leave it to steep depends on how deep of a pink you desire. I usually leave it to steep for between 6 and 9 hours. Then rinse your items in cool water and hang them to dry. •

OUTDOOR SOLUTIONS

Let's now look at some larger-scale projects that you could explore in your garden, patio or windowsill.

COMPOSTING 101

Composting food scraps is probably my most-loved actionable step to take at home. We can actually redirect the scraps from our fruit and vegetables away from landfill and turn them into powerful fertilizer that can help to grow more food. What was once deemed 'waste' can become energy.

Essentially, composting is when you create an environment that enables materials to aerobically decompose and break down to become nutrient-dense soil. Imagine an apple falling on the earth – it will slowly decompose over time and return to the soil again, right? With composting, we are emulating this process and also speeding it up by creating an environment that has heat from the sun, access to oxygen (which it doesn't receive in landfill) and our time to tend to it. Once you have set up your compost system, all it takes is a little maintenance in the form of turning, which allows the compost to breathe. Then, in four to six months, depending on how much food waste you have, you will be left with finished compost for your plants.

There are many ways to compost:

- ⊙ Build your own at-home composting system.

- ⊙ Find a local community garden that has a composting system.

- ⊙ If your community doesn't have one, pool together with friends and create one.

> Your local council might accept food scraps and organics in your garden waste curbside collection, or they might offer a local drop-off where you can take your food scraps.

> Use a private company that collects compost, which is a great option for businesses.

No matter how or where the composting takes place, you will see a significant decrease in the amount of waste you are throwing in the rubbish each week.

Getting Started

It took me a while to build the confidence to start composting at home. I was nervous that I wouldn't do it correctly, but with a little push and pep talk, reminding myself that there's no right or wrong way to do anything, I finally got

started. If you opt for an at-home composting system, this can be anything from a mound in a corner of your garden to a shop-bought composting bin or barrel. They can also be made out of an old rubbish bin, a large bucket or a stack of tyres. It is all about getting enough sun and air and balancing of food waste with drier organic material such as leaves. Learning how to compost successfully is a process, so be patient.

First things first: what can and can't be composted in your at-home compost system? The following table may be different if you are composting through another system such as your local council. Check what they accept.

CAN COMPOST	CAN'T COMPOST
Any pieces of raw or cooked vegetable	Animal products: meat and fish, bones, fat
Seeds and stones	Dairy
Coffee grounds	Packaging
Any pieces of raw or cooked fruit	Fruit and vegetable sticky labels and rubber bands
Nail clippings	Pet or human waste
Biodegradable dental floss	Weeds that have gone to seed
Human and animal hair	Ashes from grills or fireplace
Eggshells	
Untreated paper	

Most of the kitchen scraps you put into your composting area are what's known as 'green' materials. You must also balance this with what is called 'brown' materials, which are items such as dried leaves, twigs or paper (the drier, the better). Here is the difference between the two.

GREEN MATERIALS	BROWN MATERIALS
Vegetable and fruit trimmings	Brown/fallen leaves
Seeds and stones	Twigs and sticks
Stalks	Paper/cardboard that is untreated and not dyed
Garden cuttings from plants	Sawdust
Coffee grounds	Wood chips/tree prunings
Green/fresh leaves	Straw
Grass	

Here are a few troubleshooting pointers as you are getting intimate with your compost brewing:

- ⊙ If your compost is looking too wet or attracting too many flies, that means you need more brown waste.
- ⊙ If it's looking too dry, then you have added too much brown waste.

It is important that you place your compost area in as much direct sunlight as possible to speed up the process. Compost also needs turning. If you have a barrel composter (which I do), this is easier, as you can literally turn it; if you are doing more of a mound or layered bin, you will need to turn with a shovel or pitchfork. Both methods require turning; how frequently you turn it can depend on how much time you have. If you do it every few days, it will become compost much quicker than if you turn it every week. As I said, compost takes a while to master and understand, but it's a fascinating form of 'cooking', so to speak.

Knowing when your compost is ready will vary based on the type of composting system you are using. The easiest way to tell is by looking at it and feeling it. When the compost starts to resemble dark, rich soil and no longer has the ghosts of vegetable shapes appearing within it, it's nearly ready. Allow a few more weeks for it to cure and micro-organisms to stabilize. At this point, do not add anymore new compostable materials but continue to turn the pile and keep it moist. Keep a side bin or new pile for materials that accumulate during this period.

Then use a 1cm (½in) mesh screen to filter out the large chunks that can go back into the compost pile or be used as a top mulch. The screened compost can be mixed into your garden soil as an amendment for the next season, topped off as a dressing for your potted plants or mixed with regular sandy soil and coconut coir for a potting mix.

I find that composting is an enriching way to have a closer relationship to the life cycle of the food that we rely so much on as our source of energy. By taking responsibility over my food waste, I feel a tangible connection to change and solution-driven activism.

An afternoon with Manju Kumar in her garden.

MANJU KUMAR
Urban farmer and president of Brush with Bamboo

Why is creating a relationship to the soil and nature vital for our regeneration as a society?

Relationships build understanding, understanding gives us knowledge and then the right actions follow. To understand our planet, we need to be in an intimate relationship that interacts with the elements, the seasons, the natural cycles of our planet. The nurturing and honouring of soil, plants and animals deepens this understanding and connection. The easiest way for us to start is in our own gardens, patios or any space we can grow and nurture. My understanding of soil and earth is not so much from books as it is from being out there and working the land. Each year new understandings deepen my relationship with the planet and myself. Simple acts of mulching, composting and throwing seeds have helped me understand what creates a healthy garden and why I'm an essential part of the environment. We all come from agrarian roots; connecting to plants and soil is part of our innate nature. We cannot abandon our food system into the hands of a few.

What piece of actionable advice would you give the readers?

After you read this book, make sure you don't sit on the information, because it's only when we start working through the information and actually have faith in ourselves that we can do this. Read the book, then think of what steps you can take to start the journey towards revitalizing, healing, renewal.

What instills hope in you?

The fact that every day is a new beginning. The sun rises daily, the cycle of life continues, darkness is always followed by light, compost turns into rich soil; it's the nature of Earth to renew. This planet is our biggest teacher; if you want to understand something, just ask how will nature accomplish this? All the examples are in our ecosystem. If you look at a tree in the autumn, all the leaves are gone and it appears dead, but in the springtime, it blooms, life renews. Life needs moments of rest; it's a cycle of energy flowing inwards and outwards. Nature is always giving us examples of hope because it's saying there's no death; death is a transformation, it's a cycle – and cycles are just new beginnings.

Mini Herb Garden

4 plant pots, 15–30cm (6–12in) in diameter, with drain holes

1.5-cubic-foot bag of potting soil, organic, if possible, or your homemade potting mix

Hand trowel

Sage, coriander, mint and dill, organic, if possible

Spray bottle

There is nothing nicer than reaching out of your window, across your kitchen worktop, or stepping into your back garden to snip off a few mint sprigs to make tea or a small handful of dill to garnish a soup. I hope to inspire everyone to create some sort of herb garden. However, the access we each have to space is variable, and plants require love, attention and time, so it is best to start small, practise your green thumb and expand from there if you enjoy the process. I have been overzealous far too many times, ending up with failed gardening projects and dead plants, so I have to constantly remind myself that growing less, well, is better than trying out too many things at once.

A nice way to begin thinking about what to grow is choosing your four favourite herbs. If you could grow just these four items, that would be a wonderful, delicious and replenishing supply to add to your kitchen. Do some research to see if the herbs you have chosen like particular growing environments. My chosen herbs are sage, coriander, mint and dill. You will find that these herbs appear quite frequently in the 'go cook' chapter (see page 115).

I have opted to sow my seeds directly in the pots I am going to use rather than sowing them in a seed-starting tray. If you live in a particularly cold climate, starting seeds indoors will be beneficial. For seeds to grow into plants, they first need to germinate, and to do that, they need moisture, air and warmth. You can keep this entire growing process inside, as long as the pots can be placed by a window with direct sunlight and circulation of air.

In early spring, fill the pots with potting soil to about a 2.5cm (1in) from the top. Don't press too firmly; you want to leave some pockets of air within the soil. Plant about six seeds per pot – which sounds like a lot, but some may not make it. Cover the seeds with about 1cm (½in) more soil, and lightly and evenly moisten the soil with water. A spray bottle is a more controlled way to water your herbs at first; when

// continued

the plants are bigger, you can move up to a watering can or a garden hose.

With time and even moisture, your seeds will begin to grow. They could take anywhere between 2–4 weeks to sprout and rear their heads to say hello, then anywhere from 3–12 months to be harvestable. This will vary a lot depending on the herb, the weather and their care. When your herbs have reached at least 15cm (6in) in height and look healthy, they should be ready to harvest. A good rule of thumb is to only ever take one-third of the plant. If you cut too much at once, you can shock the plant and weaken it.

I hope you enjoy incorporating some of your home-grown herbs into my recipes. By growing herbs, I have a more direct relationship with the flavours in my meal. Once you have begun to master herb growing, you may feel inspired to step up to larger edible vegetables in the plant kingdom. •

○ ○ ○

Just as mending and tending to your possessions can be an act of resistance and reflection, I find making things yourself can be, too. Even the small act of making my own toothpaste or growing a plant can help me feel empowered and more self-sufficient. The items that we use every day in our home environment are incredibly intimate and personal, and I can't think of a better way to celebrate this relationship than by making them myself.

go enjoy

EXERCISES TO CULTIVATE JOY AND
CONNECT WITH THE PLANET

My intention for this book is to explore the balance between tangible, forceful action to help fight climate devastation and gentle care for the self and the planet. The hard and the soft. The love and the rage. The rest and the resistance. The crisis of the climate is heartbreaking as we are mourning the erosion of our home; it's also enraging because we have done this to ourselves. But when we allow time to be present fully to the experience of life on Earth, I truly believe we are able to see more clearly the beauty and the bounty that we still have and notice how it can be protected. It is in these moments of active rest that I feel hopeful, as I witness my interconnectedness to living things and know what a collective force we are.

In this chapter, I explore a variety of activities I enjoy that bring me closer to myself and my relatives in nature. These range from moving my body to avoid stagnation and cultivate energy, breathing practices to alleviate the anxiety of the day to day, to setting a goal of picking up five pieces of rubbish on my walk home. As someone who can fall into the fight-or-flight mode of modern-day life, I find these exercises provide me an opportunity to catch my breath.

The relationship between mind and body can at times mirror humanity's relationship to the planet, one that is deeply biologically tied but can easily lose alignment. As we realize that our planet is not a limitless resource, it is important to see that neither are we. Energy, ideas and action cannot be extracted 24/7. There must also be time to engage in nourishing activities.

When we hear phrases such as 'connecting to nature', we often think of hiking in the mountains or swimming in the ocean. But the truth is we are as much a part of nature as the mountain we wish to climb or the ocean we wish to swim in. Access to green spaces and beaches depends on many variables, but before going outside, journeying to nature starts with yourself. That is our most intimate connection. Fostering this intimate relationship with the self requires an intention to slow down and to observe and witness our own mind and body. That way we can nurture our own ecosystem so it can thrive in the outside world.

I love that our relationships with nature are so uniquely different and intimate. I encourage you to take these practices, and any practices you learn outside of this book, and make them your own.

STRETCHING AND BREATHING

I find that different types of stretching and breathing exercises can produce different types of feelings. I love to do a series of stretches when I feel anxious or frustrated, to mix up the chemicals in my body and let go. I also do certain types of breathing exercises for energy and clarity, to help me refocus on my work. Taking just five minutes out to actively breathe and stretch can help me in moments when I feel a little stuck or stagnant.

BREATHE LIKE A WHALE

Blue whales have the largest lung capacity of all living species. They produce an unmistakable blow as they exhale, which is incredibly strong and fast. About 90 per cent of the oxygen they breathe goes directly to their bloodstream, and when they exhale, about 90 per cent of the air empties out. Let's take a leaf out of their book and get present with the strength of our lungs and their ability to clear the mind and body. The one breathing exercise I never seem to forget over time is alternate-nostril breathing, a yogic breath control practice known as *nadi*

shodhana pranayama in Sanskrit. I love to do this when I begin to feel my breath shortening either from running around too much in my day or feeling anxious or before speaking at a public event.

Here are the basic steps of alternate-nostril breathing:

- ⊗ Sit in a comfortable position with your legs crossed.

- ⊗ Place your left hand on your left knee.

- ⊗ Lift your right hand up towards your nose.

- ⊗ Exhale completely and then use your right thumb to close your right nostril.

- ⊗ Inhale through your left nostril and then close the left nostril with your fingers.

- ⊗ Open the right nostril and exhale through this side.

- ⊗ Inhale through the right nostril and then close this nostril.

- ⊗ Open the left nostril and exhale through the left side.

- ⊗ This is one cycle.

- ⊗ Continue for up to 5 minutes.

- ⊗ Always complete the practice by finishing with an exhale on the left side.

STRETCH LIKE A DOG

I might spend too much time with my dog to be using this as an analogy! Dogs, like most animals, have the ability to be present, which is an image I like to take into my yoga practice. I mean, 'downward dog' says it all. Another thing I have witnessed in my own dog, Billy Blue, is his full use of space and body – he takes up as much space as he can, running, jumping and stretching into every corner of his body. I like to take these two ideas into my stretching practice: presence and taking up space.

I like to do the following stretches in the morning or at a moment in the day when I need some energy.

① Come down to your hands and knees and do a few cycles of cat and cow.

CAT: pull your belly in and round your back, letting your head hang.

COW: drop your belly towards the floor and open the chest as you look out and up.

② Still on your hands and knees, come to a neutral, flat spine and rotate your hands at the wrists so your fingers face you. Now move with a circular motion so that you are releasing your wrists, which also relieves tightness in the collar bone.

③ Rotate your hands and flip them back the normal way so your fingers are facing forwards, and extend the right leg out to the right, so the right leg is long and the foot is in line with the right hip. Now move your body in circular motions, forwards and back to open up your right hip. Once you feel open, do it on the left side.

④ Now move to a downward dog, bringing your hips up and back, feet and hands firmly on the floor, arms straight and legs slightly bent or straight, pressing back through the shoulders. Bend one knee at a time to stretch the back of the legs and hips in a pedalling motion. Once you feel open, place your left foot just a footprint in front of where it is, until you feel a stretch down the back of your left leg. Make sure to pull the left hip back and stay aligned. Take the left foot back and do the same on the right, stepping a foot-length forwards, and feel the stretch down the right leg. Return the right foot.

⑤ Walk your hands all the way back to your feet so you are in a forward fold, with bent knees. Touch your toes if you can and relax your neck. Take a few cycles of breath. Slowly roll all the way up to stand, one vertebra at a time, taking a few cycles of breath at the top before entering back into your day.

GETTING OUTSIDE

How, where and in what ways we love to move our bodies can vary significantly. No matter what it looks and feels like for each of us, moving our bodies out of the habitual positions that we work and spend our day-to-day lives in can help avoid stagnation in the body and therefore the mind.

I do love exercise, but beyond anything, I love to walk. I am certain I get my love of walking from my mum, Sheila. As a child, I was always so confused why she would want to walk home from work instead of taking the tube, extending a twenty-minute journey to nearly an hour. But I now see it was her time and space to think and transition into home life. I think the phrase 'Walk it off' is pretty accurate. If I've had a bad day or my body feels sluggish, I find the act of walking can help clear negative or unwanted thoughts and feelings. I have also lived on my own a lot in my life, and I find getting outside and watching others walk by can totally shift my mood and perspective.

TIPS FOR WALKING WITH INTENTION

- Look up and around yourself; you never know what you might see or witness.

- Set an intention – this could be to smile to a stranger walking by, practise plant and animal identification or walk a new route you haven't taken before.

- Leave your phone in your pocket, bag or even at home to resist the urge to look at it instead of your surrounding environment.

- Ask a friend to walk with you, or stop by a friend's or neighbour's to see how they are.

- On the days you don't feel like going outside, open up your windows if you can and walk around inside with the intention of shifting energy and getting fresh air.

- If you see a fallen leaf, can you take it home to draw?

- Set an intention to pick up five to ten pieces of rubbish you come across. Seeing rubbish can trigger a whole spectrum of emotions, from sadness to frustration. I pick up a piece and then immediately ask myself, What impact is this one piece of rubbish going to make? There are thousands of pieces! Which is why I give myself a manageable number.

WALKING

Walking (or what is more commonly known in America, where I am living now, as hiking) is an activity I love to do and am fortunate to have access to. My love for walking started in a very urbanized area of London where I grew up, so you definitely don't need soil under your feet to 'hike'. If you don't love to walk alone or don't feel safe doing so, ask a friend to join you or find a hiking or walking group in your area. If you'd love to check out an area that is a little less accessible, pool together with some friends and share the travel costs.

To be more respectful as I hike, these are some practices I have learned to be aware of:

> ⊙ Keep on the designated trails, as they have been created so that the majority of the landscape and its habitats can stay somewhat undisturbed.

> ⊙ Take only memories and not objects out of their habitat.

> ⊙ Leave only footprints behind and never rubbish.

The areas we hike on may be named national parks, landmarks or wildernesses, but it is important to look into the history of the areas you are enjoying recreationally. If you are in North America, like I am, we are hiking on land that belongs to Indigenous peoples. Many Indigenous nations call this land Turtle Island, a land mass that was previously subdivided into nations and tribes who each governed and stewarded their area of land. You can find out whose land you are on via native-land.ca, a website that spans the entire globe. I currently reside in Santa Monica, which is the land of the Tongva, Kizh and Chumash nations. Beyond knowing whose land I am on, I ask myself how I can be more present and mindful of the history of the lands I enjoy through hiking.

JORDAN MARIE DANIELS

Citizen of Kul Wicasa Oyate (Lower Brule Sioux Tribe), professional runner and founder of Rising Hearts

How can we cultivate a more mindful relationship to the nations of Turtle Island and America's history?

We need to start learning from the voices who have passed down this knowledge and these stories through our families, through generations. Read books by Indigenous and Black authors to understand the experience even more and to really learn about how this has been covered up or romanticized by the colonial systems at play. That's something that I ask anybody, non-Indigenous, non-Black, to just step outside of their own comfort zone and to really reflect and to sit and listen to what we're seeing, what is clearly visible on our screens, on our TV, on our phones, on everything, and to really peel back all the layers, to look at the foundation of this country and really call it for what it is: that these lands were stolen.

What is one piece of actionable advice you would give?

The biggest thing is that I'm all about self-care and healing and wellness. That within these spaces, as we fight for justice across all platforms, you need to create that intentional space for yourself to be able to decompress, to digest, to feel all the emotions and everything that comes with it, and to really set boundaries for yourself, to know your limits within these spaces as an advocate, as someone who is new to being an advocate, or a community organizer. You need to just really take care of your emotional, your spiritual, your physical well-being within this space because we need you. We need you every single day. We need that to be sustainable.

What instills hope in you?

Our next generations, especially our younger ones that are taking action and being the ones behind the mic or the bullhorn or being the loudest voice in the room to really speak to these issues. They're the ones really pushing that intersectional approach of bringing communities together and needing to work together that really inspires me, that motivates me every single day. I see them doing so much and sacrificing so much, but also I see that joy and I see their smiles, and I hear that passion and that protectiveness within them for all people on our planet. They are the ones that give me hope. They're the ones that keep me going every single day.

Hiking with Jordan Marie Daniels on Micqanaqa'n, Chumash, Tongva, Fernandeño Tataviam and Kizh land in California.

My experience hiking with Jordan and our many conversations have inspired me to be more intentional with both my hiking and my approach to my well-being as it relates to activism. I like to think how each step on my hike can be for someone else – a person, a nation, another species. This new practice has deepened my experience and relationship to the environment I am hiking in. I may go on a hike on my own, but I am never alone. I am accompanied by the memories and stories of those who have walked this path before me. I am sharing space with the animal and plant relatives around me.

Inclusion and Representation in the Outdoors and Beyond!

Many people can feel unwelcome and unsafe in outdoor activities such as hiking, camping or outdoor sports. This has been perpetuated by outdoor industries and organizations by targeting a specific type of person, both in their marketing campaigns and for staff. This is beginning to shift, but changing the faces of campaigns is not enough. Leadership roles within these companies and all industries need to be more representative, hiring and including BIPOC, LGBTQIA+ and people with disabilities.

PATTIE GONIA
Drag queen, community organizer and environmentalist

Why is it important that we change the narrative around connecting to nature?

Changing this narrative is one of the biggest keys to solving climate change. Currently, we operate in a way that separates us into two different spheres – us as humans on one side and nature on the other. When really, we are not separate worlds – we, as humans, *are* a part of nature, too. So really, connecting to nature means connecting to ourselves, to our bodies, to our unique identities and to each other just as much as it is connecting to a park or the ocean.

I also think we need to redefine and reclaim what it means to get outdoors. When we hear the word 'outdoors', we think of scenes that capitalism has tried to sell us of hyper-masculinity, grit and accomplishment. I see the outdoors as a place for self-expression, self-love, connection to each other as diverse humans and, most of all, our beautiful femininity. So go out on a trail and dance, be with friends in a garden, open a window, listen to the birds and cook yummy food. Connecting to nature will be the very soil that you will build your climate activism on. Don't forget how important it will be to tend to that soil and let's let roots run deep.

What is one piece of actionable advice you would give?

Get out there in whatever way is unique to you, and fall in love with a tree, with your identity or with other people. When you fall in love with nature, you're going to be even more equipped to fight for our planet and the people on it. Never forget, we fight for what we love.

What instills hope in you?

When I think about how I experience hope and joy in my daily life, it's *always* through connection with other people. Connection to other people often leads to collaboration on projects that add collective force to the climate movement. My activism would be a lot less effective if I was doing this work alone. Lastly, on a personal note, what gives me hope is the self-love I have finally given myself permission to feel. Growing up Queer in rural America, nature was often weaponized against me. The church I was raised in and the people around me told me that Queer people were completely unnatural and I was a flawed human. Leaving this toxic framework behind and replacing it with deep connection to Queer community and nature has shown me just how natural and frequent queerness is.

FORAGING FOR FOOD

The first time I would have foraged for food was as a child, picking wild blackberries behind our family's house on the south coast of England. There was a thick area of blackberry bushes between the sand dunes and the local park. To me, they encapsulate the taste of late summer into autumn. If my brother, Lewis, and I, with our purple-stained fingers and lips, didn't eat them all by the time we got home, we'd bake an apple and blackberry crumble.

At the time I didn't know this was a form of foraging. Foraging is the act of searching for and finding wild edible foods as a source of nourishment. Wild animals are foraging for food all the time, and we as a species used to as well. But the more domesticated and urbanized we have become, the less we have relied on found and foraged foods and the more we have relied on a food supply chain, where food must be bought. If we all had equal access to grow our own food and forage, we wouldn't need to be so dependent on our inequitable food system. This model would be defined as a food sovereignty system, where people who grow, distribute and consume the food also have a role in how the system and its policies are set up. By practising foraging in small ways, we can cultivate a closer relationship with our food sources and reframe our access to this basic human need.

INDY OFFICINALIS
Urban farmer and forager

What led you into foraging, and how has this practice connected you to the earth and nature systems? Do you see it as a form of liberation from our food system?

I was led to foraging because of a general distrust of our traditional agricultural system. I took a class on organic foods and realized how many pesticides are in our foods, and not just pesticides, but how unsustainable our food system was for people. I realized that people are heavily exploited for their labour, and I didn't want to be a part of that system. I started taking classes around just finding wild herbs and plant identification and realized that this was a really incredible way for the folks in the area that I was living in, which is a really rural area, to get healthy, highly nutrient-dense food without having to spend much money, if any money. I realized that it was a form of food liberation because people are able to get more in touch with the land and more in touch with the foods that are growing seasonally, which are typically better for your body, and also straying away from this Western food system that isn't quite nice to the people or the planet.

What is one piece of actionable advice you would give?

I would say my one piece of advice is to save seeds. I think that saving seeds from food that you eat, or even seeds that friends give you, is such a great small act of not only preservation but also of resistance. There's always this idea when you're growing up that the seeds that are in your food aren't great, or you shouldn't swallow them, or you should spit them out or throw them away. It's incredible to know that you can actually take the things that nourish you and use them to nourish other people. We're always thinking about not ourselves but also the next generation. I think it's really important to save seeds just for that purpose alone.

What instills hope in you?

Definitely the resilience of unhoused people. I am always so blown away by folks who live in a tent and just live on the street and are still very invested in their own healthcare and well-being. They're very vulnerable to a system that not only doesn't want them to take care of their basic human needs but isn't very concerned with helping them holistically heal themselves. A lot of folks that I've had the privilege of working with are really interested in ways to heal themselves that are more natural and less harsh on the environment. I feel like if folks who aren't even housed are concerned with the environment, then it should be the top priority for people who do have access to more resources and aren't in as much of a fight-or-flight mode to really care about our planet.

After my conversation with Indy, I was curious to find out which edible plants I could have access to in my area and where to start on how to forage. I am still in the early days of my foraging adventures, but here is what I have been doing so far:

- ⊘ Studying a map of my local area to understand the topography of my neighbourhood.

- ⊘ Researching the relevant laws and regulations in my area.

- ⊘ Checking which local plants are endangered and which are poisonous.

- ⊘ Reading a book on foraging specific to my area. Indy's favourite book to recommend for residents of Los Angeles is *The Urban Forager: Culinary Exploring & Cooking on L.A.'s Eastside* by Elisa Callow. *Wild Food: A Complete Guide for Foragers* by Roger Phillips is an excellent resource for foraging in the UK.

- ⊘ Keeping a journal of what I find and notice as the seasons progress or as I stumble across new and varying habitats with different characteristics.

- ⊘ Avoiding harvesting by roadsides and heavily toxic areas; the plants can absorb the toxins, and they may be present in the soil, too.

I hope these thoughts inspire a sense of curiosity in you to see if you could forage in your local area. I find just the simple act of picking a berry can not only take me back to my childhood but also connect me to the deeply ancient and primal history of our species.

GOING OFFLINE

During the course of writing this book, I have found immense peace, newfound focus and inspiration by putting my thoughts and gaze to paper instead of solely on online platforms. I love the internet and technology for the people and ideas it has connected me to, but the writing process has shown me just how important it is to take action and time offline.

SHOULDER SCREEN RELIEF

I doubt I am alone in admitting that I round my shoulders and neck, which is exacerbated through phone use, sitting and often my mood. I notice my posture most when I find myself going down an internet spiral of depressing news. Before I curl into a ball and wish the world would swallow me whole, I put my phone down and do these shoulder exercises.

① I reach my arms up and interlace my fingers, palms facing the ceiling, arms straight. I then stretch my arms side to side, bending my elbows, wrapping either inner elbow around my head. This is a nice shoulder flossing exercise that gets into all the corners of your shoulder joints.

② With your left hand, grab on to your right wrist and guide it up and overhead to the left. Angle your face and gaze to under your right armpit. For a deeper stretch, if you are doing this exercise standing, lift your left heel up and bend at the waist further over to your left, feeling a deep elongating stretch down your right-side body. Now alternate sides, with your right hand grabbing on to your left wrist, and lean over to the right, feeling the left-side body stretch out.

③ Sit down with your legs out in front of you. Take your hands behind your back and rest the tops of your hands on your mid-back, fingers interlaced. Now slowly lie backwards onto your arms; they will act as a guide to encourage your shoulders open and down. This is such a nice deep opening in the shoulders and the front of your chest. The easiest way to get out of this position is to bend your knees, press into the feet to lift your hips, then your arms can easily unfold and separate under your back.

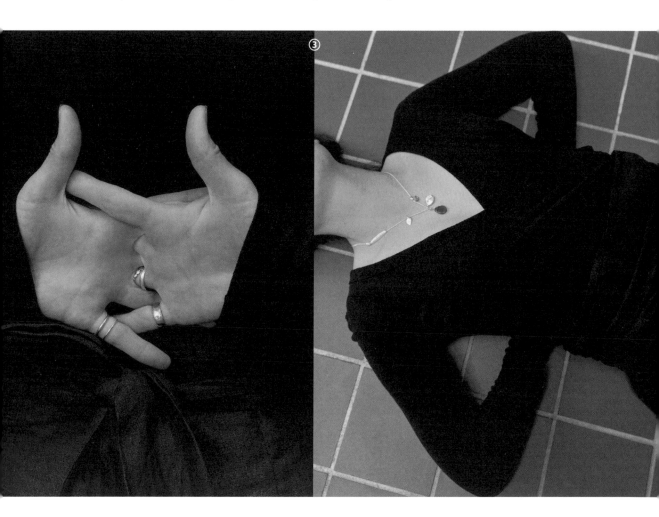

SOCIAL MEDIA/NEWS DETOX

I have to admit I am not very good at avoiding social media and news. But writing that down here and being transparent is inspiring me to be better at it. What I *am* good at is not looking at my phone when I first wake up, which hugely helps my mood in the morning, my favourite time of day. But I see more and more people online and among friends practising total social media and news detox either over the weekend, on their workdays or at changeable times depending on their schedule.

I have found it easiest to start with a couple of hours or set a time to go offline, say from 6:00 pm till I wake up. Or on the weekend, not going online till say 4:00 pm in the afternoon. You choose the number of hours to have a little detox; but once you decide, really try to stick to it. It is a great practice of boundaries, and when you achieve your tech detox you will feel accomplished and proud of yourself for sticking to it. I always feel calmer after I take these breaks, as my relationship to my phone and other devices feels healthier.

JOURNALLING

There are definitely environments such as the internet to load us up with a dictionary of mixed emotions, and there often isn't a tangible space to make sense of how we are feeling. Quick text exchanges between friends is an opportunity for sharing feelings back and forth, but we can fall into the trap of answering 'How are you?' with 'Okay' because we honestly just don't know how we feel. Living in a time where the climate crisis and social justice issues are talked about daily and we've experienced a global pandemic, anxiety can be high. I am often reminding myself, or more often being reminded by loved ones, that 'it's okay to not be okay'. Journalling can help us make sense of the 'not okay' by going deeper and unpacking why we feel that way.

It took me a while to get the hang of it, but I've learned to love the journalling process. I find that by sitting down with a bit of paper or a notebook and letting my thoughts out on the page without overanalysing, I'm able to really get in touch with how I'm feeling. I love the visual aid of a Feelings Wheel, which has helped me develop a deeper vocabulary to express my emotions.

Here are some question prompts that I like to work with. I hope they can support you, too.

Today I felt most disconnected to myself when _____ happened, and it made me feel like _____.

I read an article or a social media post today about _____, and it made me feel _____, which was confusing because _____.

I worked hard shifting _____ habit within the home and found it challenging because _____, which made me feel like this _____.

Someone shared their opinion on the climate crisis; my reaction was _____, and it made me feel _____.

I felt most at peace and calm today when _____, and it felt _____.

I witnessed or read an uplifting story about _____, and it made me feel _____.

Today I am setting myself the small task of _____, and I hope it makes me feel _____.

Now that you have the feelings wheel, try to set aside some time each day to journal; maybe it's five minutes, maybe it's twenty. Maybe it happens when you first wake up in the morning, as a reset in the middle of your day, or to wind down right before bed. I can go in and out of using journalling as a tool and instead turn to breaking a sweat with exercise or chatting to a friend to understand how I feel. The important thing to remember is that this, like all of the exercises in this chapter, is an opportunity for reflection and rest. Don't worry about what you write, how you move or what you say, just start.

go beyond

FINDING YOUR PLACE
IN THE CLIMATE MOVEMENT

I hope by now that this book has offered you real, tangible solutions and ways to participate in the individual change you can achieve from the intimate environment of your home. Now that the foundational concrete is beginning to set, it's time to look at the actions you can take beyond the home with your wider community, branching out from friends, neighbours, colleagues, peers, groups and folk around the globe. You may have noticed that different topics within this book resonated more with you; this could be a wonderful starting point as you look at your role within social change.

There is a wide and beautiful spectrum of roles within social change, not just the ones we see on the front covers of newspapers, magazines or in our Instagram feeds. While we each may have our own image of a stereotypical activist or advocate, that mould is there for us to break and redesign. The climate justice movement needs each and every one of us to show up imperfectly and truthfully as who we are.

MIKAELA LOACH

Medic, climate justice advocate, cohost of the *YIKES* podcast and claimant on the Paid to Pollute court case

How did you become an activist?

Growing up, I did things that we were told to do like fund-raise for charities, lend my time to organizations and volunteer where I could. Then, as I got into my late teens, I got involved with migrant rights work and refugee rights advocacy, and I also became aware of my lifestyle choices and actions and how that impacted the planet and people. But I still saw activists as strong, confident, flashy people who stood up on podiums, gave speeches and went to protests. I didn't see myself in that, but I also wanted to create change. When I went to Calais the first time, I saw people who were activists in a different way than I'd seen before. People who put ego aside and just did the work that needs to be done; the work that might be a bit more boring and might not get as much praise. These real-life activists showed me that, one, I could be an activist, too; and two, I had already been doing activism without really knowing it.

What piece of actionable advice would you give?

Just join in a group with other people. So often we can get in our own heads about change and making change and action, and make it all about us, and what can we do alone. Actually, when you join a community with other people, when you connect to other people, you realize that not only are you not alone but also when you work for people, when you're with other people, you learn so much more.

What instills hope in you?

There's a quote from Arundhati Roy that says, 'Another world is not only possible, she is on her way. On a quiet day, I can hear her breathing'. When I'm thinking about hope, I just try and think where can I hear that breath? When I feel that breath the most is when I'm in the community with other people or when I see other people in a community working together, unifying around a cause to try and create a better world. I think hope is action. We can be that hope. We can be that breath. I think that shows the power that we can have together. That's what keeps me going.

When I cofounded Waste Watch with my friend Georgia Stockwell, we were desperate to get involved in a climate action network in our neighbourhood. We didn't want to sit quietly reading about the climate crisis anymore, unsure of how to take action and feeling like we were only learning from the internet and not people. So we started our own group with about six people coming together on my living room floor, which then evolved to about a seventy-person group that met bimonthly. At each event, Georgia and I moderated panels and conversations on particular topics around the climate crisis, with the intention to learn as a group and inspire action. During one of our meetings, I learned that you can go to your city hall and speak in favour or against a proposed bill. So I did that and spoke in favour of a bill that would require LAX airport and the city concessionaires to include plant-based protein options on their menu. I can't tell you how great it felt when it passed!

Georgia Stockwell and me setting up for one of our Waste Watch meetings in Los Angeles.

Through Waste Watch, I experienced how essential it is to go beyond our individual actions, as it allows us to connect with other people and action opportunities. If we only focus on our home environment, it can distract us from the issues at large and from addressing the climate crisis with the urgency that it needs addressing. Any moment I catch myself feeling overwhelmed, confused or hopeless, I always look to friends and community for guidance and to lift me out of my story before letting it freeze me in inaction. We are stronger together and wiser by listening to each other.

The movement is made up of so many different people, each with different lived experiences and things to give. There are also so many facets to the climate movement. Like any ecosystem, community needs biodiversity to thrive. So, before we worry that we are not enough, don't know enough or don't have the 'right' experience, it is important to remember that this movement needs all of us. The big question always is, 'Where do I start?' The journey to finding the right role for you can be full of ups and downs. I believe we each have a calling that feels unique to us and plays to our strengths and sensitivities, and through experimentation and curiosity, we can find it. The activities in this chapter get to the core of your individual strengths and provide you with tools and encouragement to get involved in the wider community. It's important to stress right here that you know yourself best; my suggestions are just a loose framework to help guide you to what is already within you.

REFLECTING ON YOUR VALUES

To begin this journey of finding your role within social change, it's useful to first reflect on your values as an individual and notice where and what you are attracted to.

Ikigai, 'The Reason You Get Up in the Morning'

Ikigai, best translated as 'the reason you get up in the morning', is a Japanese term to describe that sweet spot of purpose, the intersection of what brings you joy, what you are good at and what the world needs. Feeding into this could also be your past experience, vocation or skills, perspective and identity. I think when we stop to reflect on *ikigai,* we can see beauty in what we each individually have to offer.

A great exercise is to write a couple of bullet points for each of the outer three circles. I like to come back and do this at times when I feel a little distracted and need re-centring on my path and purpose. You will begin to notice how they intersect and hopefully spark thought for further inquiry. Then to expand on this, here are some questions I often ask myself to cultivate more clarity:

- ⊚ What events and experiences happening in the world right now concern me?

- ⊚ Is there something, someone or somewhere I am passionate to protect?

- ⊚ When I contemplate the natural world and planet Earth, where do I picture myself within that? Am I currently standing alongside it, closely related or separate from it?

- ⊚ When I think about the human experience, am I thinking about my relationship to my neighbours and close community or am I picturing the experience on a more global scale?

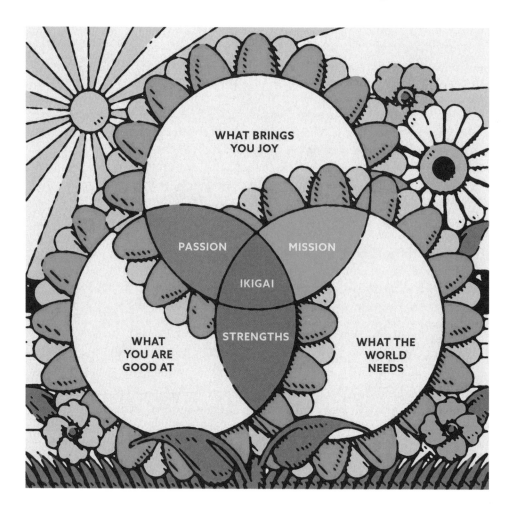

⊚ What experiences, people or environments are making me feel confident and expanded, and which can cause me to shrink and feel overwhelmed?

⊚ Is there something that I believe is holding me back from forming the relationship I want to with the climate movement? Is it a skill, confidence, access or the knowledge of where to start?

⊚ When I imagine myself within the climate movement, doing what I would love to be doing with others, how do I feel? How does this lived experience make me feel physically, emotionally, intellectually and perhaps spiritually?

I hope these questions are able to get some ideas and feelings flowing. Now let's move into possible roles within community organizing and social change.

YOUR ROLE WITHIN SOCIAL CHANGE

Behind the scenes of organizing, there are many different roles we often don't see or hear that much about. Not everyone wants to be the person behind the microphone, and there are so many equally important, if less visible, roles within the movement. Now there is no cookie cutter to fit into, so while I highlight some key role types below, I encourage you to find space between two roles or even define one for yourself.

I have tried my best to create the ultimate role identifier – a sorting hat of sorts – that will help guide you to the most authentic and exciting role within social change. It is the knowingness of what makes you tick and the ownership of your unique skills and perspective that will empower you on your journey. I have outlined five different roles that, through questioning, experimenting and keeping an open mind, I hope can bring you more clarity. This may be an instinctual choice or a challenging one, so I encourage you not to be limited by these roles. If you feel like you are a combination of two, that's great, too. If you try one and realize it's not for you, that is valuable information. I find discovering what you don't enjoy can be as beneficial as learning what you do.

The Roles

Visionaries

You notice where systems are not working and are able to imagine and design new ways of thinking and operating. You are not afraid to think and dream big. If someone hands you a blank canvas or a protest sign, you are quick to grab your pencils and get to work on filling the page. You enjoy coming up with new ideas and thinking outside the box.

Organizers

Nothing excites you more than bringing people together and facilitating change. You like to text, call, email and communicate with individuals and large groups. You feel a great sense of calm when the job gets done. People often turn to you for guidance and leadership.

In the Field

You feel most alive when you roll up your sleeves and get your hands dirty. You enjoy being at the epicentre of the movement, whether that's big crowds or grassroots local projects. You feel most fulfilled when you can witness a tangible change happening, whether that's seeing a seed you planted grow in your local community garden or being among thousands of people at a protest that makes headline news.

Behind the Scenes

You are a quiet, steady worker who likes to get the job done and be part of something bigger than yourself. You listen well and thrive as part of a team. You are a natural caregiver who likes to support others. You are perceptive and can quickly notice where something or someone needs aid. You are passionate and patient in your journey. Preferring to be behind the scenes and not centre stage, you are quiet but essential.

Storytellers

You observe people, ideas, events and movements unfold and have a unique way of communicating what you see with a wider audience. You listen well to others and have great compassion for their stories. You are able to weave a narrative that is inclusive and captivating.

Questionnaire

This questionnaire is designed to help you find a role within the five I have outlined. It is a multiple-choice questionnaire, so circle your answer with a pencil so that you can come to a conclusion at the end. If you remember those questionnaires in teen magazines where you answered questions to find out your celebrity crush or which Spice Girl you are, then you will be familiar with this format.

Before you go to sleep, which of these do you think about?

a. Endless ideas; it's sometimes hard to sleep.
b. What you have to get done tomorrow.
c. Watering your plants.
d. Calling a family member or friend.
e. Someone you met that day and the stories they told you.

You see a listing for a volunteering opportunity. Which of the following time commitments are you interested in?

a. The role is mobile, flexible and can be discussed on application.

b. A six-week full-time commitment to a campaign.

c. A single event with a time and a date.

d. Once a week for a minimum of one month, with the opportunity for more if interested.

e. A weekend seminar/convention event.

Which of the following headlines grab your attention?

a. An environmental after-school-club initiative or organization loses its funding.

b. Your town's summer festival gets postponed.

c. The oldest tree in your local town is cut down to make way for a new shopping centre.

d. The nurses union is in a dispute with a local hospital about overtime pay.

e. The board of a film festival again fails to elect a diverse jury.

Which of the following experiences most interests you?

a. You meet someone you look up to.

b. A group text with friends makes plans to meet up.

c. You meet a new friend at a local community event.

d. You catch up with a neighbour you haven't seen in a while.

e. You hear a talk by an author at your local bookshop.

If there's a conflict at work, you are more likely to . . .

a. Start plotting how to quit and start your own business.

b. Send an email to colleagues to arrange a staff meeting.

c. Approach the issue or person directly.

d. Keep it to yourself unless it happens again.

e. Recount the story over dinner with your friends.

If you are watching a film with a friend, you will be the one . . .

a. Noticing all of the continuity mistakes.

b. Asking your friend what their plans are at the weekend and if they want to hang out.

c. Stretching on the floor and finding it difficult to sit still.

d. Quietly watching.

e. Commenting on characters and plot.

A framed picture fell off the wall – you are more likely to . . .

a. Decide it was time to rearrange all the pictures on the wall.
b. Ask the handy person in your house to address is.
c. Immediately get the toolbox out and rehang it.
d. Assess the damage to the photo and check all other hanging photos are secure.
e. Stare at the photo, recall memories and forget to rehang it.

If you answered mostly As, you're a visionary; mostly Bs, an organizer; mostly Cs, in the field; mostly Ds, behind the scenes; and mostly Es, a storyteller. Remember, you can be a combination of two. This is just a fun exercise, and it might not work for everyone. Now the next big step is joining a community within the movement.

HOW TO FIND YOUR COMMUNITY WITHIN THE MOVEMENT

Getting involved in the climate movement by joining a community group is an amazing action you can take on your journey to protect and fight for our future on this planet. The moment we immerse ourselves in community, we are more likely to commit long term, find joy in our action and be empowered to keep persisting. This could be in person or virtually, with one other or a hundred others, depending on your personality or circumstances.

Joining a community might not necessarily mean finding a new one; first consider the communities you already belong to. These could be your friend group; as a parent, it could be with other parents; if you play sports, this could be your teammates; people you share a hobby with; or your family, classmates, place of work or worship. The friends you have in these communities might also be interested in integrating climate issues and action into your group. For instance, if you play football, your team might be interested in starting a park clean-up group in your local area, or friends at work might want to create a sustainability initiative within the company. And if they are not yet interested, this could be an opportunity to suggest it.

If you are seeking a fresh new group to expand your circle and learn something new outside your current communities, here are some questions to ask yourself as you begin that search:

⊙ Am I looking for a weekly commitment, participating in a one-off campaign or event, or taking a few actions with the same group a few times a year?

⊙ Do I want to meet up with a community in-person or online?

⊙ Do I have a particular focus and intersection within the climate movement I am interested in?

⊙ Is my identity (your culture, sexuality, ethnicity, beliefs or values) a key factor in what I am looking for in a community?

⊙ What would I like to give to the community?

⊙ What would I like to learn from the community?

⊙ What types of actions would I like to take part in?

⊙ What scale of group would I like to join? A large nonprofit, a local chapter of a large nonprofit, a small organization in my area, a small organization with community members around the country or globe?

⊙ Would I like to have a leader or member of authority leading the group, or would I prefer a more leaderless, co-operative model of a community?

⊙ Am I interested in action focused on the neighbourhood or city I live in or more global issues?

It is also useful to ask yourself what type of community you *don't* want to be a part of. I find if I am stuck on navigating what I do want, it's often easier to first refine boundaries of what I am not looking for.

Once you have some key themes and requirements you are looking for, it's time to start finding your group. It may sound obvious, but the first step could be as simple as researching online either through a search engine or platforms such as Instagram, Facebook or LinkedIn.

This search could look a little like these:

'climate group near me'

'community garden volunteering in (name of your town)'

'my Greenpeace local chapter'

'queer youth climate group'

When my concern for the environment and humanitarian issues began to grow about ten years ago, I had no idea what to focus on within these huge issues. I felt lost and lacking in direction. Yet the closer I observed my feelings and curiosities, the more the ocean kept popping up as a theme. So I started to go to talks on ocean health, read books about it, follow organizations and try to meet people in this field. It was a light bulb moment when I met ocean campaigners and scientists who worked with Greenpeace. It became clear to me that this was my path.

I share this because at this point in your journey, you may not yet have a specific interest, so it might be good to start with a larger organization with many different pathways. This is a great way to learn a bit about the climate issues at large, so you can figure out which areas are most attractive to you. There is no one size fits all; we each have different curiosities, ways we would like to be involved and types of groups we want to be a part of. Now you may be super lucky and find your community quickly, but it may take a few attempts, showing up to meet-ups, signing up for newsletters, reading forums and emailing groups. Each time you reach out and put yourself out there, it is guiding you to the right place. You might meet someone at one community event, such as a climate rally, and they lead you to another community, and then another, and that's when you finally find a space that feels like home. There is no rush, and there is no one right way or perfect group. Keep being curious and vulnerable in your search. Remember, people are looking to find you as much as you are looking for them. Sounds a little like relationship advice, but I really think community is our greatest relationship.

Performing deckhand duties with fellow Greenpeace crew on the *Arctic Sunrise* in the Gulf of Mexico.

MARQUISE STILLWELL

Founder and principal of Openbox and cofounder of Urban Ocean Labs

How can collaboration and storytelling, from your experience as a filmmaker and designer, be used to get people involved in the climate movement?

Storytelling is so important when it comes to actually getting people to understand the changes that need to happen, the way that we need to live our life and the way that we need to reimagine. Film is very powerful. It sucks you in, and all of a sudden, you're in a fairy tale. It's filled with make-believe. Once it's made, people really believe certain aspects of a story when it comes to fiction. How might we leverage that when it comes to telling other stories? The intersection of storytelling, of film, of design, along with policy and science to help confirm the rigour of the thinking around the story is so important.

What is one piece of actionable advice you would give?

For me, it begins with modelling. It could be as simple as picking up paper when you walk down the street. You're picking up paper because there's rubbish and it needs to be thrown away, but you're also modelling behaviour. Human beings are driven by behaviour and we're driven by model behaviour. If people just throw rubbish, then everyone's going to throw rubbish. If people start picking up the rubbish, people are going to start to feel like, 'Oh, that's what I should be doing'. It's the simple steps that help to create behavioural change.

What instills hope in you?

To me, hope is about being human. To be human is to be a part of a large ecosystem, and we need to be responsible. My hope is in responsibility for not only my individual actions, but it's the power of the 'we'. I'm the greatest (we) I have ever known. I can't do anything by myself. My hope is in the power of the we.

INTEGRATION

I have been privileged that my interest in the climate movement and my integration of positive habits in my day-to-day life have been, for the most part, accepted by the people I am around. That's not to say I haven't had people roll their eyes and say, 'You really think that's going to make a difference?' But in the grand scheme of things, I recognize I have not been treated differently or bullied because of my participation in the climate movement. If you have experienced this or you are worried you will be as you begin to participate in the climate movement, then here are some ways to integrate this important part of you into your existing relationships. These key relationships could be with your family, roommates, colleagues, partner, friends or classmates.

Channel Confidence and Don't Make Assumptions

As I learned about the climate crisis and wanted to start implementing new habits into my life, I was a mix of overly enthusiastic to share my feelings with all my friends and then kind of hesitant, feeling like it was easier in some ways to keep it to myself. In hindsight, I wish I had been more confident, but it can be difficult to ignore the fear of being judged. It's also important not to assume that other people aren't interested in taking action. Say, for instance, you think your place of work isn't dealing with its waste in the most mindful way. Rather than striding up to your boss and shaming the company, it is better to first ask if there are any plans to create a better waste system in the office. If not, mention that you have some ideas and would be interested in talking about it.

Use Inspiration as a Tool

When it comes to sharing new habits, I believe that the best way to inspire others is to continue as you are, embodying these new things and expressing the joy they bring you. Allow people to be inspired and approach change in their own time. This, of course, doesn't mean hiding all of these new exciting things away for no one to see; it's just a reminder to be gentle with how you share them. This could be the difference of allowing a friend to compliment you on your reusable coffee cup rather than saying something like, 'Oh, don't you have a reusable yet?' We can't assume this friend hasn't already got their eye on a cup they're saving up for or wondering if a mason jar would be good. We are each on our own schedule of implementing change in our daily lives.

Stay Calm and Stand Your Ground

If and when you do come up against friction with others, it can be frustrating and upsetting. But before treating fire with more fire, I have found it best to listen and calmly, coolly have a conversation. It's important to hear someone out even if their opinions are opposite to yours. Something I have experienced is people wanting to push my buttons by questioning my scientific and statistical knowledge of the climate crisis. The truth is, I definitely do not know everything. I am not an expert; I am just an enthusiast. Don't let people doubt you just because you are on a journey of learning. Sometimes you just have to stand your ground, have the courage of your convictions and agree to disagree.

SOCIAL PROOF

Human beings love to follow the crowd. As much as we like to think we are all uniquely individual, a lot of what we do is based on the actions of others. When these are positive things, this art of persuasion and ability to create new trends are incredible tools in scaling our actions. This is why it's important to integrate our new habits and actions both inside our homes and into the public domain. When we do things in public spaces – such as supermarkets, on public transport or at the gym – we are on the best possible stage to influence others and create waves of change. When people witness other people doing things, we are statistically more likely to do them. This is even more likely if the person doing them is respected. This is called social proof, and it's a tool that many marketing teams use to sell products and services.

A good example of this could be at the supermarket checkout when you are packing your shopping into your reusable bag and the person behind you in the line notices and could think to themselves, *Ooh, I must remember my bag next time*, and they do. When I think about the concept of social proof, it encourages me to keep living my values out in the open. I will never be able to see my impact on others, but knowing that I create a little ripple in the ocean that could one day join other ripples to become a wave is a powerful reminder for me.

ACTION

The last thing I want you to do is read this book and sit on the information. Of course, we all take our own time to digest and reflect when we have read a book, and I recognize that on a grand scale, the amount of work required to address our climate issues is immense and can feel completely overwhelming. But we must not be frozen in inaction. Whenever I feel like my individual or community actions are not enough, I think about one of the founding principles of Greenpeace: the idea of bearing witness. As Annie Leonard, executive director of Greenpeace USA, explained to me, 'The idea of bearing witness is that we have a responsibility to know what's going on. Then we say, once you know, you owe. Once you've seen harm happen, you have a responsibility to at least tell other people, but preferably even get involved'. I remind myself that now that I know, I owe. As informed humans with voices and platforms and opinions, it is our joint responsibility to question, share and, most importantly, take action. Once you start, whatever that may look like for you, I guarantee you will be so much more inspired and encouraged by your actions, community and progress than you will be discouraged by the magnitude of the crisis.

From left to right: Talking with media aboard Greenpeace's ship the *Arctic Sunrise*. At a climate rally with youth climate activists Kevin J Patel and Alexandria Villaseñor.

FINALLY, THIS BOOK IS JUST THE BEGINNING, NOT THE END

I wrote this book from my own experience of becoming involved in the climate movement. I was called to share with you the intimate and individual actions I was taking quietly at home in order to have a more tangible relationship to the climate crisis. This journey has led me down many paths; some have split into two, some have been dead ends and some I have turned around on. During the process of writing this book, the most important thing I have learned is that our work does not end with our individual actions, it only begins here. Like you, I am still at the beginning. I have a lifetime of commitment to this movement and so much more to learn and do. I hope this book builds a foundation and gives you the tools to take on bigger challenges that exist outside the home and the self.

The Earth – and that includes the human species – needs us to show up with the collective in mind. We are no longer about systems that only serve the 1 per cent, we are about a fair and just system that gives everyone a seat at the growing table. We are co-conspirators in designing the future we want. I think one of the greatest gifts of the human species is our ability to imagine. So let's use our radical imagination to not only change our perceptions of the climate crisis but also to change our conditions. I believe we have agency over how we wish to leave this planet for our ancestors.

Thank you for being part of this book and the ecosystem that we as a community are building within our beautiful biosphere, Planet Earth.

go gently

RESOURCES

Books

Brandt, Christen, and Tammy Tibbetts. *Impact: A Step-by-Step Plan to Create the World You Want to Live In*. New York: PublicAffairs, 2020.

Callow, Elisa. *The Urban Forager: Culinary Exploring & Cooking on L.A.'s Eastside*. Altadena: Prospect Park Books, 2019.

Cameron, Julia. *The Artist's Way: A Spiritual Guide to Higher Creativity*. New York: TarcherPerigee, 1992.

Carson, Rachel. *Silent Spring*. Boston: Houghton Mifflin, 1962.

Figueres, Christiana, and Tom Rivett-Carnac. *The Future We Choose: The Stubborn Optimist's Guide to the Climate Crisis*. New York: Vintage Books, 2020.

Hanh, Thich Nhat. *How to Walk*. Berkeley: Parallax Press, 2015.

Harari, Yuval Noah. *Sapiens: A Brief History of Humankind*. London: Harvill Secker, 2014.

Hawken, Paul. *Drawdown: The Most Comprehensive Plan Ever Proposed to Reverse Global Warming*. New York: Penguin Books, 2017.

Johnson, Ayana Elizabeth, and Katharine Wilkinson. *All We Can Save: Truth, Courage and Solutions for the Climate Crisis*. New York: One World, 2020.

Kimmerer, Robin Wall. *Braiding Sweetgrass: Indigenous Wisdom, Scientific Knowledge, and the Teachings of Plants*. Minneapolis: Milkweed Editions, 2013.

Klein, Naomi. *On Fire: The (Burning) Case for a Green New Deal*. New York: Simon & Schuster, 2019.

Klein, Naomi. *This Changes Everything: Capitalism vs. the Climate*. New York: Simon & Schuster, 2014.

Ligotti, Thomas. The *Conspiracy against the Human Race: A Contrivance of Horror*. New York: Penguin Random House, 2018.

Macfarlane, Robert. *Underland: A Deep Time Journey*. New York: W W Norton, 2019.

Martin, Molly. *The Art of Repair: Mindful Mending: How to Stitch Old Things to New Life*. London: Short Books, 2021.

Morton, Timothy. *Dark Ecology: For a Logic of Future Coexistence*. New York: Columbia University Press, 2016.

Schlossberg, Tatiana. *Inconspicuous Consumption: The Environmental Impact You Don't Know You Have*. New York: Grand Central Publishing, 2019.

Sheldrake, Merlin. *Entangled Life: How Fungi Make Our Worlds, Change Our Minds & Shape Our Futures*. New York: Random House, 2020.

Shiva, Vandana. *Soil not Oil: Environmental Justice in an Age of Climate Crisis*. Berkeley: North Atlantic Books, 2015.

Solnit, Rebecca. *Hope in the Dark: Untold Histories, Wild Possibilities*. New York: Nation Books, 2004.

Wallace-Wells, David. *The Uninhabitable Earth: Life After Warming*. New York: Tim Duggan Books, 2019.

Weisman, Alan. *World Without Us*. New York: Picador, 2007.

Wenders, Wim, and Mary Zournazi. *Inventing Peace: A Dialogue on Perception*. London: I B Tauris, 2013.

Williams, Terry Tempest. *Erosion: Essays of Undoing*. New York: Sarah Crichton Books, 2019.

On screen

An Inconvenient Truth. Directed by Davis Guggenheim. Los Angeles: Lawrence Bender Productions, 2006. Documentary.

Anthropocene: The Human Epoch. Directed by Jennifer Baichwal, Edward Burtynsky and Nicholas de Pencier. Ottawa: Mercury Films, 2018. Documentary.

Before the Flood. Directed by Fisher Stevens. Los Angeles: RatPac Documentary Films, 2016. Documentary.

Frontline. Season 38, episode 16, 'Plastic Wars'. Directed by Brent E. Huffman, Katerina Monemvassitis and Rick Young. Aired March 31, 2020, on PBS. TV episode.

Gather. Directed by Sanjay Rawal. Brooklyn: Illumine Group, 2020. Documentary.

Ice on Fire. Directed by Leila Conners. Los Angeles: Appian Way, 2020. Documentary.

Kiss the Ground. Directed by Joshua Tickell and Rebecca Harrell Tickell. Santa Monica: Benenson Productions, 2020. Documentary.

The Repair Shop. Directed by Ben Borland, Craig Ellis and Emma Walsh. Aired March 27, 2017. Documentary, reality-TV.

RiverBlue. Directed by David McIlvride and Roger Williams. Vancouver: Paddle Productions, 2017. Documentary.

Seaspiracy. Directed by Ali Tabrizi. Santa Rosa: Aum Films, 2021. Documentary.

Story of Plastic. Directed by Deia Schlosberg. New York: Pale Blue Dot Media, 2019. Documentary.

Audio

For the Wild. Ayana Young. September 30, 2014. https://forthewild.world/about-us. Podcast.

How to Save a Planet. Kendra Pierre-Louis, Rachel Waldholz and Anna Ladd. July 31, 2020. https://gimletmedia.com/shows/howtosaveaplanet/episodes#show-tab-picker. Podcast.

Yikes. Mikaela Loach and Jo Becker. February 11, 2020. https://podcasts.apple.com/us/podcast/the-yikes-podcast/id1498623503. Podcast.

Drilled. Amy Westervelt. November 14, 2017. https://drillednews.com/podcast-2/. Podcast.

Emergence Magazine Podcast. Emergence Magazine. February 6, 2018. https://emergencemagazine.org/podcast/. Podcast.

Yale Climate Connections. Anthony Leiserowitz. August 9, 2014. https://yaleclimateconnections.org/the-climate-connections-podcast/. Podcast.

Websites

gogently.earth

www.greenpeace.org

www.atmos.earth

www.intersectionalenvironmentalist.com

www.rainforestalliance.org

www.choose.love

www.Native-land.ca

www.raceforward.org

www.ifixit.com

www.futureearthcatalog.com

www.slowfactory.foundation

www.terracycle.com

https://www.gov.uk/guidance/organic-food-uk-approved-control-bodies

https://ofgorganic.org/about

www.fairtrade.net

https://www.soilassociation.org/certification/

www.fsc.org

www.regenorganic.org

www.mcsuk.org/goodfishguide/

https://www.rspcaassured.org.uk/about-us/

REFERENCES

go learn

Stephanie Shepherd, Max Moninan, (@futureearth). '@mary.heglar for the soul', Instagram photo, August 1, 2020.

Sean Fleming, '3 Billion People Could Live in Places as Hot as the Sahara by 2070 Unless We Tackle Climate Change', World Economic Forum, May 13, 2020, https://www.weforum.org/agenda/2020/05/temperature-climate-change-greenhouse-gas-niche-emissions-hot/.

Abrahm Lustgarten, 'The Great Climate Migration', *New York Times Magazine*, accessed February 2, 2021, https://www.nytimes.com/interactive/2020/07/23/magazine/climate-migration.html.

'Renewable Energy', Center for Climate and Energy Solutions, accessed February 23, 2021, https://www.c2es.org/content/renewable-energy/.

'World Atlas of Desertification', Joint Research Centre, European Commission, accessed April 25, 2021, https://wad.jrc.ec.europa.eu/introduction.

Emma Charlton, 'You Eat a Credit Card's Worth of Plastic a Week, Research Says', World Economic Forum, June 24, 2019, https://www.weforum.org/agenda/2019/06/you-eat-a-credit-card-s-worth-of-plastic-a-week-research-says/.

'How Much Oxygen Comes from the Ocean?' National Oceanic and Atmospheric Administration, accessed on April 2, 2021, https://oceanservice.noaa.gov/facts/ocean-oxygen.html.

Bethan C O'Leary et al., 'Effective Coverage Targets for Ocean Protection', *A Journal for the Society for Conservation Biology* 9, no. 6 (November/December 2016): 398–404, http://onlinelibrary.wiley.com/doi/10.1111/conl.12247/epdf.

'National Overview: Facts and Figures on Materials, Waste and Recycling', Environmental Protection Agency, accessed on January 28, 2021, https://www.epa.gov/facts-and-figures-about-materials-waste-and-recycling/national-overview-facts-and-figures-materials.

'What a Waste 2.0: A Global Snapshot of Solid Waste Management to 2050', The World Bank, https://datatopics.worldbank.org/what-a-waste/trends_in_solid_waste_management.htm

Paula Dutko, Michele Ver Ploeg and Tracey Farrigan, 'Characteristics and Influential Factors of Food Deserts', United States Department of Agriculture, no. 140 (August 2012): 1–2. https://www.ers.usda.gov/webdocs/publications/45014/30940_err140.pdf.

Christina Nunez, 'Climate 101: Deforestation', National Geographic, February 7, 2019, https://www.nationalgeographic.com/environment/article/deforestation.

Sarah Gibbens, 'Less Than 3 Percent of the Ocean Is "Highly Protected"', National Geographic, September 25, 2019, https://www.nationalgeographic.com/environment/article/paper-parks-undermine-marine-protected-areas.

'Regulating Fisheries Subsidies', United Nations Conference on Trade and Development, accessed on April 4, 2021, https://unctad.org search?keys=Regulating+Fisheries+Subsidies.

'AR5: Climate Change 2013: The Physical Science Basis', Working Groups, Intergovernmental Panel on Climate Change, accessed on April 16, 2021, https://www.ipcc.ch/site/assets/uploads/2018/02/ar4-wg1-chapter5–1.pdf.

'Hyperion Water Reclamation Plant,' LA Sanitation & Environment City of Los Angeles, accessed on February 22, 2021, https://www.lacitysan.org/.

'What a Waste: An Updated Look into the Future of Solid Waste Management', The World Bank, September 20, 2018, https://www.worldbank.org/en/news/immersive-story/2018/09/20/what-a-waste-an-updated-look-into-the-future-of-solid-waste-management.

'Environmental & Climate Justice', National Association for the Advancement of Coloured People, accessed on April 10, 2021, https://naacp.org/know-issues/environmental-climate-justice.

'The Success of Nonviolent Civil Resistance', International Center on Nonviolent Conflict, accessed on March 2, 2021, https://www.nonviolent-conflict.org/resource/success-nonviolent-civil-resistance.

'Global Trends: Forced Displacement in 2020', UNHCR The UN Refugee Agency, accessed March 21, 2021, https://www.unhcr.org/flagship-reports/globaltrends.

Saeed Kamali Dehghan, 'Climate Disasters "Caused More Internal Displacement Than War" in 2020', The Guardian, May 20, 2021, https://www.theguardian.com/global-development/2021/may/20/climate-disasters-caused-more-internal-displacement-than-war-in-2020.

Gleb Raygorodetsky, 'Indigenous Peoples Defend Earth's Biodiversity – But They Are in Danger', National Geographic, November 16, 2018, https://www.nationalgeographic.com/environment/article/can-indigenous-land-stewardship-protect-biodiversity-.

'Indigenous Peoples', Understanding Poverty, The World Bank, accessed on April 5, 2021, https://www.worldbank.org/en/topic/indigenouspeoples.

Enrique Salmón, 'Kincentric Ecology: Indigenous Perceptions of the Human–Nature Relationship', Ecological Society of America 10, no.5 (October 2000): 1327–1332, https://www.jstor.org/stable/2641288.

go see

'Food Loss and Food Waste', Food and Agriculture Organization of the United Nations, https://www.fao.org/food-loss-and-food-waste/flw-data)

'How Much Do Our Wardrobes Cost to the Environment', The World Bank, September 23, 2019, https://www.worldbank.org/en/news/feature/2019/09/23/costo-moda-medio-ambiente.

Sabbira Chauduri, 'The Tiny Plastics in Your Clothes Are Becoming a Big Problem', *The Wall Street Journal*, Updated March 7, 2019, https://www.wsj.com/articles/the-tiny-plastics-in-your-clothes-are-becoming-a-big-problem-11551963601.

'2021 Resale Report', ThredUp, accessed on March 22, 2021, https://www.thredup.com/resale/#resale-industry.

'Extending Clothing Life Protocol', Waste and Resources Action Programme, accessed on March 22, 2021, https://wrap.org.uk/resources/guide/extending-clothing-life-protocol.

Jessica Kane, 'Here's How Much a Woman's Period Will Cost Her over a Lifetime', HuffPost, Updated December 6, 2017, https://www.huffpost.com/entry/period-cost-lifetime_n_7258780.

Hannah Brooks Olsen, 'How Much Does a Period Cost, Anyway?' Medium, October 16, 2017, https://medium.com/s/bloody-hell/how-much-does-a-period-cost-anyway-6a2263828ae3.

Allison Sadlier, 'New Research Reveals How Much the Average Woman Spends Per Month on Menstrual Products', SWNS digital, November 27, 2019, https://swnsdigital.com/2019/11/new-research-reveals-how-much-the-average-woman-spends-per-month-on-menstrual-products/.

'Toilet Paper History', Toilet Paper History, accessed March 2, 2021, http://www.toiletpaperhistory.net/.

Kristen Hall-Geisler, 'Is It Scratchy? 5 Things to Know about Bamboo Toilet Paper', HowStuffWorks, July 13, 2021, https://home.howstuffworks.com/green-living/bamboo-toilet-paper.htm.

Sarah Griffiths, 'Why Your Internet Habits Are Not as Clean as You Think', BBC, Smart Guide to Climate Change, March 5, 2020, https://www.bbc.com/future/article/20200305-why-your-internet-habits-are-not-as-clean-as-you-think.

'This Search Engine Plants Trees', Atlas of the Future, accessed on March 18, 2021, https://atlasofthefuture.org/project/ecosia/.

go shop

Jann Bellamy, 'Food Fights in the Courtroom', Science-Based Medicine, July 24, 2014, https://sciencebasedmedicine.org/food-fights-in-the-courtroom/.

'Recycling Facts', RecyclingBins.co.uk, https://www.recyclingbins.co.uk/recycling-facts/.

The EU Ecolabel Product Catalogue, European Commission, http://ec.europa.eu/ecat/.

go keep

Jordan Wirfs-Brock and Rebecca Jacobson, 'A Watched Pot: What Is the Most Energy Efficient Way to Boil Water?' Inside Energy, February 23, 2016, http://insideenergy.org/2016/02/23/boiling-water-ieq/.

Matthew Gault, 'National Right-to-Repair Bill Filed in Congress', Vice, June 17, 2021, https://www.vice.com/en/article/v7e37d/national-right-to-repair-bill-filed-in-congress.

'Why Love Your Clothes?' Love Your Clothes, accessed on April 23, 2021, https://www.loveyourclothes.org.uk/about/why-love-your-clothes.

'Washing Laundry in Cold Water Protects a Lot More Than Just Our Clothing', Cold Water Saves, accessed April 23, 2021, https://www.coldwatersaves.org.

Madison Alcedo and Emily Belfiore, 'The 8 Best Dryer Balls to Reduce Static and Drying Time', Real Simple, April 1, 2021, https://www.realsimple.com/home-organizing/cleaning/laundry/best-dryer-balls.

'What does the new energy label mean for you and your home?', *Which?*, https://www.which.co.uk/news/2021/03/what-does-the-new-energy-label-mean-for-you-and-your-home/.

'Tap Water Can Add Up to a Big Waste', *San Diego Union Tribune*, November 6, 2015, https://www.sandiegouniontribune.com/lifestyle/sdut-saving-water-brushing-teeth-2015nov06-story.html.

'Residential Toilets', Water Sense, Environmental Protection Agency, accessed on April 1, 2021, https://www.epa.gov/watersense/residential-toilets.

go make

Victoria Allen, 'Wet Wipes Could Take 100 Years to Break Down: Products Contain Plastic That Is "Virtually Indestructible"', *Daily Mail*, November 6, 2016, https://www.dailymail.co.uk/news/article-3911606/Wet-wipes-100-years-break-Products-contain-plastic-virtually-indestructible.html.

'Facts', CoffeeSock, March 19, 2021, https://coffeesock.com/facts.

go enjoy

'Which Animal Has the Largest Lungs?' Reference, March 28, 2020, https://www.reference.com/science/animal-largest-lungs-41bff7885213a7a7.

'The Anatomy of a Blue Whale', Whale Watch Western Australia, April 2, 2020, https://whalewatchwesternaustralia.com/single-post/2020/04/02/the-anatomy-of-a-blue-whale/.

ACKNOWLEDGEMENTS

I firstly want to thank all the authors before me. It is within the pages of the books I have read that I have found courage to write my own. I have never been someone who has found reading easy, but the challenge in that has driven me to read more, which then led me to writing.

Thank you to my agent, Nicole, for your unwavering understanding of my vision and integrity throughout this creative process. Thea, thank you for bringing clarity, cohesion and confidence to my writing. Thank you to my editorial team at HMH and HarperCollins. To my creative team that brought the book to life, Kacie, Kim, Amanny, Courtney, Amber, Michael and Chloe, thank you for making this book such a beautiful and lasting object.

To all the people I have had the privilege to interview for *go gently*, I am so grateful for your time, expertise and unique perspective on the climate crisis. Thank you, David, Joanna, Kevin, Magdalena, Leah, Richard, Annie, Marta, Manju, Jordan, Pattie, Indy, Mikaela and Marquise.

To my family, Sheila, Gary and Lewis, who have been physically far away from me through the writing of this book but, as always, the closest. Thank you for challenging me creatively and intellectually to make this book the best it can be.

Andrew, thank you for being there every step of the way. You have kept me calm and centred. I am so grateful for your patience and love.

Thank you to all those who care for our planet and fight for the protection of all living things. Past, present and future.

INDEX

Note: Page references in *italics* indicate photographs.